GENES, CELLS AND BRAINS

GENES, CELLS AND BRAINS

THE PROMETHEAN PROMISES OF THE NEW BIOLOGY

HILARY ROSE AND STEVEN ROSE

VERSO
London • New York

FOUNDATION
BOOKS

FOUNDATION
BOOKS

Foundation Books is an imprint of Cambridge University Press, India.

Cambridge House, 4381/4 Ansari Road, Daryaganj, New Delhi 110 002

www.cambridgeindia.org

This edition is published by arrangement with Verso Books.

First published by Verso 2012
First South Asia edition 2015

Printed in India at India Binding House, Noida

ISBN 978-93-84463-47-2 Paperback

Contents

Introduction: Prometheus Unbound?

In the Greek myth, Prometheus was the Titan who both created the first person from clay and gave humanity the fire he stole from the Gods. His punishment was to be tied to a stake, his liver devoured by an eagle during the day, only to be regrown at night. Mary Shelley drew on this myth, calling Victor Frankenstein, the creator of her monster, *The Modern Prometheus*; her Prometheus, however, was no longer a god, but a scientist. But why was Frankenstein disgusted, fleeing from the very being he created? He had set out with noble intentions, using the new science of electricity to fashion life from dead tissue. As he explains: 'In a fit of enthusiastic madness I created a rational creature, and was bound to him to assure, as far as was in my power, his happiness and well-being.' The monster tells Frankenstein the cost of rejection, of his denial of love and refusal to take moral responsibility for the monster he has created: 'There is love in me the likes of which you've never seen. There is rage in me the likes of which should never escape. If I am not satisfied in the one, I will indulge the other.'

The Promethean claims of today's fusion of biomedicine and biotechnology to explain, mend, manipulate and transform the lives of the wealthy global minority are becoming ever louder. Indeed, they have already become part and parcel of everyday assumptions and discourse. Evolutionary theory offers to explain human origins, genomics to define similarity and difference, genetic and stem cell therapies to cure or prevent disease and even enhance bodies and minds, the neurosciences to predict behaviour, to explain consciousness and, with brain organisation theory, to re-essentialise sex gender difference. Genetics, this time with benign intent if not with benign consequences, has sought to re-racialise human difference.

In the process the life sciences have been transformed into gigantic biotechnosciences, blurring the boundaries between science and technology, universities, entrepreneurial biotech companies and the major pharmaceutical companies, or 'Big Pharma'. Knowledge becomes intellectual property. The technosciences take place within and as part of a globalised economy, made possible by digitalisation, and stretching from the older scientific centres of Euro-America to the rising eastern giants of China, Singapore and India. The major players in these changes have been Big Pharma itself, venture capital, biotech companies, the state with its interests in surveillance and control, and, as usual, the military. In their train have come new and formidable powers not only to reconstruct but to construct life itself. The human cancer-bearing Oncomouse produced by Harvard biologists has become the icon of the reconfiguration of those seemingly fixed boundaries between culture and nature. DuPont holds the patent on this living creature, but Oncomouse is neither just nature nor just culture; only the neologism of culture/nature does justice to the technosciences of life in the twenty-first century, a brave new world, where life is both made and born.

In the chapters that follow, we trace the unfolding narratives of the biotechnosciences of genomics, regenerative medicine and the neurosciences – or, as in our title, Genes, Cells and Brains – locating them within the global neoliberal economy and culture of the twenty-first century. Genomics begins with the most ambitious and costly project in the history of the life sciences, the sequencing of the DNA of the human genome. Even as the international Human Genome Project began in the 1990s, plans were being laid for the creation of massive DNA databanks, intended to link the health records of entire populations with their DNA, to identify 'disease genes' and develop personalised medicine. Hopes that this would lead to breakthroughs in the form of new drugs foundered on the gene sequencers' failure to recognise the sheer complexity of humans as biosocial creatures, shaped by both evolutionary and social history. Nor has it been easy for genomics to transcend its own history, above all its inextricable links with eugenics. As genetic hopes diminished, a new prospect emerged: the almost magical potential for human embryonic stem cells to enable the lame to walk and the blind to see. And after stem cells has come the promise of the neurosciences as therapy to cure ill and disordered minds and as culture to construct human identity. From Genes'R'us to Neurons'R'us in two decades.

This is new territory, with new threats and promises. Our own generation grew up in the boom years of full employment and the security provided by the welfare state, but also under the shadow of the Bomb. Back then, the possibility of nuclear war – whose genetic effects, assuming we were spared total annihilation, would stretch down through the generations – gave a sense of precarity to this apparently secure everyday life.

The tumultuous year of 1956 saw both Khrushchev's secret speech denouncing Stalinism and the Soviet invasion of Hungary, which together triggered the greatest crisis in Western communism; in Britain alone, ten thousand members left the Party. In that same year came the Suez crisis: Britain and France, the old imperialist powers, in collusion with Israel as the last of the white colonisers, invaded Egypt. They had badly miscalculated the public reaction. In Britain there was an immediate and outraged response not just from the left, the trade unions and the churches – even the right wing of the Labour Party came out in opposition. Out of this ferment was born a New Left searching for a new kind of politics. New Left Clubs sprang up as places to debate ideas, rather than expound the correct line. The same willingness to experiment saw the New Left develop an alliance with the pacifists, out of which grew the Campaign for Nuclear Disarmament (CND) – the first of the new social movements.

It was as part of this new movement, with its debates, demonstrations and marches, that we (the authors) first met at the London New Left Club, in the unlikely premises of the nightclub at 100 Oxford Street. Social and natural scientists rarely write together, even though both have to grapple with complexity and contingency. Nonetheless the epistemological gap needs minding, and the hierarchy of the sciences is a constant irritant. Perhaps even more complicated in our case is that that we live together in a heterosexual relationship, and despite the gains of feminism are all too conscious that the 'tradition of the dead generations weighs like a nightmare on the brain of the living'.

That we share a commitment to social justice and democracy, have been influenced by Marxist thought, taken part in many social and cultural struggles and between us debated the relationship of the study of bios, as life itself, and the study of the social, led her quite early on to acknowledge the materiality of the body and him the reality of the social. Despite this it was only after we had published our first joint book, *Science and Society*, in 1969,[1] that we realised we should have known better – the book should have been called *Science in Society and Society in Science*. 'Science' is not separable from 'society' but part of it.

WHO BENEFITS?

Marx famously asked: *Cui Bono*? In the depths of the current crisis, the question of who are the political and economic beneficiaries of capitalism and now of the capitalist technosciences resounds louder than ever. For the bankers, the 1 per cent, the answer is clear. But for the 99 per cent? Just who are the main beneficiaries of the Human Genome Project, the massive DNA biobanks, stem cell research and the great expansion of the neurosciences?

Science – knowledge about the physical and biological world – was once seen as independent of the society and culture within and as part of which it was generated. Marx and Engels themselves both saw science as a progressive force within society and at the same time recognised the knowledge it produced as reflecting the interests and ideology of the capitalist class. This analysis was taken up with enthusiasm in the infant Soviet Union, and in 1931 Boris Hessen, a member of the Soviet delegation to the International Congress of the History of Science in London, electrified a new generation of young scientists already radicalised by the suffering inflicted on the working classes by the Depression with a challenge to one of the cornerstones of modern physics. His paper 'The Socio-economic Roots of Newton's Principia', argued that this most arcane of mathematical treatises and the physics it formalised developed in response to the needs of the rising seventeenth-century mercantilist capitalism.[2]

Among those inspired by Hessen was the crystallographer and polymath Desmond Bernal, whose *Social Functions of Science*, published in 1939, became the foundational text for the social relations of science movement.[3] Bernal believed that science was in itself socially progressive but had been perverted within the capitalist mode of production. A true science for the working classes – a proletarian science, and one which fully liberated its potential – could only be realised within socialism.

This vision of a proletarian science faded in the 1940s when, in the Soviet Union, the fraudulent agronomist Trofim Lysenko's attack on 'bourgeois genetics' was backed by Stalin, ending in the destruction of Soviet genetics (and of many geneticists).[4] With this entry of the Cold War into the laboratory, Bernal, like other communist scientists in the West, retreated to the conservative ideology of science as a neutral pursuit, to be used or abused by society. Instead turning his political attention to the gross imbalance between the research budget allocated to the military and that allocated to civilian ends, he began a long campaign of 'science for peace'. This doesn't mean that Bernal altogether abandoned the theoretical question of the social relations of science – his finest and least positivist contribution was yet to come, with his 1952 pamphlet *Marx and Science*.[5]

THE RISE OF THE RADICAL SCIENCE MOVEMENT

It was the global opposition to the Vietnam War that saw a renewed questioning of the political identity of science. One strand in the anti-war movement included those biologists morally outraged that their discipline was being recruited into a monstrous war against a poor peasant society. Research on plant hormones had been re-engineered by military scientists to produce chemical defoliants, new eco-genocidic weapons directed against the forests, crops and people of Vietnam.[6] By the end of the 1960s, the biologists' anger over what they saw as the 'misuse' of their science boiled over in the pages of the leading scientific journals, in campus teach-ins, laboratory occupations (spectacularly in Japan and Italy) and demonstrations on the streets. Out of the anger came a new radical science movement, for some part of the counterculture, for others part of the New Left. The new movement challenged the ideology of the neutrality of science, calling for its democratisation and the building of a science for the people.

The movement was fluid, always linked internationally, but shaped by its national context. In Italy, a strong Marxist tradition meant that the flamboyant activism of the laboratory occupations was matched by the movement's theoretical strength. Physicist Marcello Cini (part of the leadership of the New Left Manifesto group) in his critique of the commodification of science provoked a response from philosopher Giovanni Berlinguer, brother of Enrico, the leader of the Italian Communist Party.[7] The same issues were debated by the libertarian groups, Lotta Continua and Potere Operaio.

French scientists were not far behind; during the May events of 1968 the Paris laboratories were empty. Like the Italians, the French were well versed in Marxism. Physicist Jean-Marc Lévy Leblond wrote of the double ideology – the ideology of science and the ideology in science.[8] Physicist Monique Couture-Cherki and sociologist of science Liliane Stéhelin raised the issue of sexism within science, the former criticising the exclusion of women, the latter exposing the androcentric ideology of science, in which women seeking to become scientists had to become pseudo men.[9]

By contrast, in the US and UK, few of the younger generation of science activists were familiar with Marxism or the history of the social relations of science movement. In the US the opposition to President Johnson's 1965 escalation of the Vietnam War began with the California-based Scientists and Engineers for Social and Political Action and the East Coast Science for the People. Students campaigned against the host of military contracts with the universities, and discovered that what Eisenhower had termed the military-industrial complex was now a military-industrial-scientific complex into which US universities were inextricably locked. Biologists were important in the struggle against scientific racism, with the collectively written pamphlet *Sociobiology as a Social Weapon* providing a powerful rebuttal.[10] Individuals such as the palaeontologist Stephen Jay Gould and the geneticist Richard Lewontin, both Marxists, were leading biologists and brilliant polemicists.

Meanwhile the global women's liberation movement was sharpening the consciousness of women scientists, particularly biologists already involved in the radical science movement as writers and activists. These campaigns attacked the institutionalised discrimination against women in science and the damaging cultural claims made by bad and biased biology that women were in nature the inferior sex. Psychologist Ethel Tobach, molecular biologist Rita Arditti, biochemist Ruth Hubbard, and physiologist Ruth Bleier published foundational texts challenging patriarchal science.[11]

In 1969 two young and politically active molecular biologists opened a new front, this time concerning the environmental and human risks posed by advances in molecular biology. The two, Harvard molecular geneticists Jon Beckwith and James Shapiro, were senior authors of a paper in *Nature* reporting the first ever isolation of a gene (the bacterial *lac operon*).[12] But instead of proudly hailing their scientific and technical triumph, Beckwith and Shapiro used the occasion to call attention to the hazards that the research carried with it, in particular the possibility of genetic manipulation by modifying DNA (recombinant DNA), and the risk of the escape of transformed bacteria into the environment with unforeseeable consequences for plants, animals and humans.

Molecular genetics, they claimed, offered modern society unprecedented power to manipulate bios. In Promethean mode this could be for human benefit (and profit) and in sceptical mode a dangerous threat. Beckwith and Shapiro's warning thus fed into a slowly building wave of public anxiety concerning the risks posed by the new biotechnology. In response, the National Institutes of Health (NIH) established a Recombinant DNA Advisory Committee, swiftly copied in the UK by the Wilson government's Genetic Manipulation Advisory Group. The City Council of Cambridge, home of Harvard, had heard enough; they called for the banning of such research within the city limits. In the face of this public alarm and hostility, leading molecular biologists became increasingly worried – though whether they were concerned about the risk to the environment or more about the future of molecular biology was far from clear.

As a result, in 1974, Paul Berg (soon to win a Nobel Prize for his DNA research) called together a unique conference of genetic researchers at Asilomar, in California. The conference proposed a voluntary moratorium on genetic manipulation, and guidelines on containment facilities to prevent possible escape. The conference was also seized on as an opportunity for the scientists to discuss the commercial potential of the research. This was followed up in 1976 with the formation of the first of the new Californian biotech companies, Genentech, by another recombinant DNA Nobelist present at Asilomar, Herbert Boyer, together with the venture capitalist Robert Swanson. The era of geneticist-entrepreneurs had dawned.

In Britain too, the first stirring of the radical science movement developed out of opposition to the Vietnam War. By 1967 biologists, including Steven (Rose), were speaking at campus anti-war meetings attacking the US military's eco-genocidic defoliants and the lethal use of CS gas. A survey carried out by Hilary (Rose), of Vietnamese who had fled the South to escape the defoliants, drew early attention to the probability (later confirmed) that these were causing both cancers and birth deformities.[13] Another studied the effects of the British use of CS during the late '60s to control the insurrection of the Northern Irish nationalists. This drew attention to the fact that while its effects on the young able-bodied rioters were marginal, it caused considerable physical distress to the vulnerable young and old. CS gas in this civilian context was more a political own goal than a successful crowd-control technology.

The Old Left scientists of the social relations of science movement, that generation of 'science for peace', welcomed the young activists, particularly the natural scientists among them. They initially supported the newly formed British Society for Social Responsibility in Science (BSSRS), but with the movement's commitment to new non-hierarchical organisational forms and its position on the non-neutrality of science they gradually drifted away.

As in the US, by the mid-1970s UK feminists, angered at the androcentricity of the movement, broke away to form independent groups. The Brighton Women and Science collective produced *Alice Through the Microscope*,[14] the London group published a special women's issue of BSSRS's magazine *Science for People*, while yet others began the long critique of the new reproductive technologies.[15] Nonetheless the central figures in the British New Left, and above all its key journal, *New Left Review* (NLR), remained trapped in what the physicist-turned novelist C.P. Snow had famously characterised as the Two Cultures. *NLR* ignored science, both as culture and as a material force integral to capitalism's relentless pursuit of innovation. The journal's leading theoretician, Perry Anderson, dismissed Bernal's contribution as consisting of 'fantasies', 'false science', 'to be blown away by the first gust of the international gale'.[16]

The fledgling BSSRS was mostly innocent of the theoretical debates about science that had raged earlier in the century. Instead it fostered activism – campaigns on industrial hazards, IQ, safe food, pollution – and set up 'science shops' bringing local communities and scientists together. Separately, the *Radical Science Journal*, with historian of science Robert Young as its central figure, positioned itself as the theoretical journal of the movement. Young argued that science is reducible to its social relations – proposing an ontological philosophical relativism over and above historical and sociological relativism – a position fiercely challenged by critical realists in the pages of Ralph Miliband and John Savile's *Socialist Register*.[17] The new theoretical preoccupation with the social production of scientific knowledge became a central concern both within and without academia. Several of the natural and social scientists who had been involved in the radical science movement professionalised this interest, becoming part of a growing international academic community of the social studies of science and technology.

THE RISE OF THE GREENS

In 1962 Rachel Carson's book *Silent Spring* had put the impact of chemically synthesised pesticides on the environment onto the international political agenda.[18] Almost single-handedly, Carson stimulated a new social and political awareness of risk and a new Green movement. While the near global scale and flamboyance of the subsequent campaigns, as environmentalists hugged trees and tore up GM crops, compelled governments and biotech companies to act, opposition to biomedical technosciences remained relatively muted.

It was the German Greens in the 1970s who led the opposition to the geneticisation of biomedicine. They were also the most broad-based of the European Greens, including environmentalists, Marxists, feminists and the churches. For Europeans generally and the Germans in particular, the prospect of genetics being used to identify foetuses with 'abnormalities' raised the spectre of eugenics. Likewise, many feminists saw the new reproductive technologies as increasing the power of a patriarchal science and technology over women's bodies; their opposition was led globally by the Feminist International Network of Resistance to Reproductive and Genetic Engineering. The European Parliament (more progressive then than now), spurred on by the German Greens, successfully blocked the European Commission's programme on human genomics – the programme's name, Predictive Medicine, giving the game away. The Commission repackaged the project, leaving out the provocative concept of 'predictive', and by the 1990s the European Human Genome Project went through.

Three decades on from *Silent Spring*, the Weberian sociologist Ulrich Beck developed his theory of the risk society in which this new risk to nature and society replaced the old risks of social precarity, which had, in his view, been resolved by the welfare state.[19] While Beck's assumption may have held for Germany's sturdy Bismarckian welfare state it applied less well, if at all, to those countries in which the welfare state was being rolled back at speed. In these countries the new risks, rather than replacing the old, joined them. But for many sociologists the elegance of Beck's theory was so entrancing that they failed to check whether it stood up.

THE CHANGING PRODUCTION SYSTEM OF SCIENTIFIC KNOWLEDGE

In the current fusion of biomedical reductionism and techno-optimism the historical distinctions between science and technology, pure and applied science, academic, industrial and military research, today hold weakly if at all. In the biotechnosciences researchers move seamlessly between them all, as consultants, entrepreneurs, company directors and shareholders. Some could, with equal accuracy, be named capitalists as much as scientists. With this transformation in both the macro-political economy and the production processes of knowledge, the values of life scientists have changed. In the past they were largely 'disinterested', that is, they were focused on the knowledge of natural entities; they hoped for recognition and may have dreamed of Nobel Prizes, but more mundanely settled for an adequate salary and a secure pension. Making what one Nobel Prize winner spoke of enthusiastically as shed-loads of money, or becoming a celebrity scientist, getting huge advances for popular science writing, having a personal television series, having 'interests', are all part of this new world. Mammon has been welcomed into the laboratory.

This loss of disinterestedness has become a much-discussed problem in the leading scientific journals. Commercial secrets cannot be shared. Researchers have been prosecuted for transferring biological samples from one academic laboratory to another. PhD students can work for months on a project only to find that they cannot continue as they have run into a patent. Competition has weakened the once cooperative values of the academic research community. Refereeing research papers and grant applications presents new difficulties: how can disinterested refereeing be maintained when the commercial interests of the referee may conflict with those of the scientist being assessed? The journals have fought to maintain standards by insisting that authors declare conflicts of interest, but this is not easy to police.

The journals too have financial interests. Most are owned by commercial companies. The leading journal *Nature* is the property of Macmillan, and the giant Anglo-Dutch publisher Reed-Elsevier owns many hundreds of the most prestigious academic journals. *Science* turns a profit for its owner, the American Association for the Advancement of Science. University libraries, compelled to purchase these journals, sag under the costs, and a series of rivals, including the open access Public Library of Science, has been created – but here the scientists themselves have to pay to be published. As a result, scientists working in poor countries and weak institutions can now read the journals but their chances of publication in them remain weak.

The rediscovery of the significance of the norms and values of the scientific community has to some extent taken the sociology and history of science back to Robert Merton's thesis, which dominated the field in the mid-twentieth century but has been largely abandoned since the 1970s. Merton was interested in the ways in which the cultural structure of science guarantees scientific objectivity, that is, the values of science as an institution rather than the values of scientists as individuals. At the time, his view, now taken for granted, that as individuals scientists were no more or less ethical than anyone else was startling. Merton, however, saw the shared norms of the scientific community as crucial.

He identified four key values: *Communalism*, in which scientists share their findings cooperatively in exchange for recognition (Merton initially called this communism, and for rather obvious reasons soon sanitised the term); *Universalism*, ensuring that claims are evaluated in impersonal ways, setting aside questions of status such as nationality, race or religion; *Organised Scepticism*, which demands that all scientific claims must be scrutinised by the scientific community; and lastly *Disinterestedness*, or the assumption that scientists are not influenced by personal material gain; that their reward lies in recognition by their peers.

Today attitudes towards intellectual property undermine

communalism – and are in radical opposition to Merton's initial concept of communism. Universalism has come under siege from what have been variously termed the new social justice and the new identity movements. As a matter of justice these new movements have fought to bring hitherto excluded identities into the polis. Integral to this struggle has been an insistence on the significance of knowledge from below, initially understood as the knowledges arising from class, later also of gender and race. Thus standpoint theory derived from the call for a proletarian science by Marxists in the 1930s, followed much later by the '60s scientific radicals' demand for a 'science for the people', and still later in the 1980s for feminist science, black science and Islamic science.

Among the feminist standpoint theorists were American political scientist Nancy Hartsock, philosopher Sandra Harding, and British sociologist Hilary Rose.[20] They were soon followed by the African American sociologist Patricia Hill Collins, who pointed to the specific location of black feminists and hence of black feminist thought. All these 'situated knowledges', as Donna Haraway defined them, threatened the claimed universalism of the natural sciences in both its neutrality and its objectivity. Situated knowledge increases the sophistication and complexity of the humanities and social sciences but brings difficulties for the natural sciences.

Organised Scepticism remains, but with so much of the technosciences shrouded by industrial secrecy, such that data cannot be shared and discussed openly, and with Big Pharma suppressing so much negative or uncomfortable data from its drugs trials, this too is constrained. Disinterestedness, likewise, can no longer be guaranteed. Unquestionably the norms and values of the scientific community have changed, to the extent that the objectivity that Merton believed was guaranteed by the old cultural structure of science is in trouble. The historian of science Steven Shapin, in his 'moral history of a late modern vocation' (science), sees disinterestedness as retreating from academic research but still present in some private biotech laboratories.[21] Were Merton alive today, he would surely have asked what has happened to the guarantee of objectivity, when the cultural structure of science and its norms have so radically changed with the technosciences in the age of globalisation.

There was, however, a problem with the Mertonian thesis even when it was initially proposed, as his focus was on academic science and academic science alone. Academic science was and is the tip of the iceberg of science visible above the waterline; what goes on beneath is invisible and frequently the larger part. Even in the 1960s, when the Mertonian paradigm was very much in its heyday, more than 70 per cent of British research was carried out by industry or the military. A rare study, *The Scientific Worker*, exploring the norms and values within industrial laboratories (conducted by Norman Ellis, a PhD student of the historian of science Jerry Ravetz at Leeds), found the Mertonian norms in short supply. There was no Weberian sense of vocation that so preoccupies Shapin; what the industrial researchers did was a job, albeit an interesting and reasonably well paid job. These scientists were a high-skilled stratum within the proletariat.

Even this abbreviated account suggests that the question of how society gets its sciences and how they grow is a good deal messier than the classical philosophies of science would have us believe. Even as late as the 1970s Karl Popper's view of the growth of science through 'bold conjectures and refutations', based as it primarily was on a Whig history of physics, dominated the field, with its flattering image so welcome to the natural scientists. And yet this internalist view took no account of the social world in which science had both developed and been shaped. To ignore the Manhattan Project and its conclusion in the bombing of Hiroshima and Nagasaki was to turn more than a blind eye to the political dimensions of the growth of physics. In the decades following 1945, spurred on by the military demands of the new Cold War, spending by state and industry on research and development increased exponentially in the US, Europe, the USSR, and later in Japan, before levelling off by the 1980s at around 1.7–2 per cent of GDP. Throughout these long decades of growth, an appointed expertocracy controlled the science budgets and directed priorities, although a long tradition held that, despite being largely state-funded, state and government should stay at arm's length from both the universities and academic scientific research. The point here is thus not to trace the subsequent modifications and assaults on Popperian theory led by Thomas Kuhn and Paul Feyerabend, but to understand how it came to be that natural scientists – above all the physicists who required large sums of money to pursue their heroic projects – secured the critical resources. How did the demands of state, industry and the military shape the direction of the sciences?

Globalisation – industrial, financial, political, informational and cultural, with its virtual abolition of distance of space and time – is central to the growth of twentieth- and twenty-first-century biotechnology. Just as the phenomenon of a risk society arising from scientific and technological development preceded its social theorisation, so too with globalisation, as the economist Amartya Sen reminds us. Sen's timescale is longer and his range of geographical and cultural difference greater. Over the *longue durée*, he points out, much earlier globalisations arising from the East have moved to the West. But the era of America-driven globalisation is coming to an end, as Brazil, Singapore, India and China flex their technoscientific muscles. Over the past five years China has overtaken the UK and the US to run the biggest and fastest genome-sequencing industry in the world.

Genomics would have been impossible without the revolution in informatics, itself driven forward above all by the US military's search for global reach. By the end of the twentieth century there was no corner of the globe, no aspect of life that had not been caught up in the transformation made possible by the fusion of Tim Berners-Lee's civilian WorldWideWeb with the US military's Internet. (Today China competes not only in genomics but also in cyberwar). The life sciences have been empowered by this revolution to create a new hybrid form, located in a new space between the university and industry. The old disciplines of science and technology mutate and merge. Hybridity proliferates. Industrial laboratories with all their requirements for secrecy are increasingly located within the university campus itself, with the science parks built for the academics' spin-off companies handily close by.

Modern molecular biotech is utterly dependent on information technology, with its capacity to instantly compare and analyse protein and gene sequences, design chemical structures, and search papers and patents from widely dispersed centres of research. The international proliferation of DNA biobanks is underwritten by informatics, and as the biobanks need larger and larger data sources they too foster increasingly global knowledge structures. In the age of globalisation biomedical information, along with bios itself, is commodified.

Globalisation is key too to the success of Big Pharma, whose twelve largest companies have a combined annual turnover of £400 billion and a global reach, continuously relocating research activities and clinical trials. The increasing costs of clinical trials in the West and the difficulty of finding subjects who had not been treated with similar drugs has led the Pharma companies to outsource clinical trials to poor countries. This has been welcomed by Eastern European governments embracing neoliberal economics and seeking inward investment. Estonia, for example, advertises its educated and science-friendly population as ideal for drug testing. The ethical and political sensitivities associated with research on human reproduction have given a new twist to the problems of finding locations for human embryonic stem cell research. Here the alternatives are either countries with little or no regulation or those where there is clear but not too demanding ethical regulation.

The British, in their desire to lead the field technically and commercially, chose the latter path. In the US it is entirely possible for Federal and State policies to conflict. During the Bush presidency it was illegal for Federal funds to be used for human embryonic stem cell research, but States and private companies can do as they choose. California not only permitted human embryonic stem cell research but allocated more funds that the researchers themselves could usefully employ. Texas has gone further, permitting stem cell clinics, even though the Federal government refuses to license the procedures. And there is no prohibition on private research into human reproductive cloning. Despite Obama overturning Bush's prohibition, the ensuing legal fight has perpetuated the unstable environment.

DEMOCRATISING TECHNOSCIENCE?

The political project of democratising science has come in three waves. The first, the social relations of science movement of the 1930s, foundered thanks to the Cold War and Lysenko's fraudulent science. The second wave was the radical science movement of the 1960s and '70s, part of that great surge of social movements calling for democracy and accountability, rejecting any automatic deference to expertise.

Committed to the New Left and the radical science movement we believed that democratisation was possible, and that our generation had begun that long march through the institutions – including those of science and technology. To democratise science, the scientists themselves needed to be more representative of the general population – above all not to be so overwhelmingly male and white – and the processes of science policy making opened up for democratic scrutiny. The institutions, including those of science and technology, did of course change as the generation of the 1960s with its values entered them.

Feminism compelled the institutions of science and technology to include more women, and the presence of those women themselves carrying feminist values extended the process. Thus where biomedical science had previously focused on the bodies of men as the universal representatives of the human race, considering women only in the context of reproduction and psychology, the different cycling bodies of women were now studied. Women's hearts, once solely a metaphor for love and caring, now became material organs to be studied alongside those of men. Globalisation ensured that the great laboratories of Europe and America became multinational and multicultural, but without significantly opening their doors to their own citizens of colour.

The third wave began with the German Greens in the early 1980s. Genetic engineering, with GM crops as the trigger, was seen by the

Greens as the negation of the new politics they were trying to build. It's worth recalling that many of the Party's founders had been part of the 1968 generation and their rainbow alliance radically confronted Germany's Nazi past and its eugenic science. With this the Greens had put both the question of trust in science, and democracy as its answer, onto the political agenda. A point reiterated by Beck.

In 1985, aware not only of the growth of public distrust in science and with it the possible knock-on effect on research funding, London's Royal Society established a Committee on the Public Understanding of Science (COPUS) under the geneticist Sir Walter Bodmer. The Committee's stance, and the programme that it initiated, was built around the thesis that if only the public understood science better they would trust it more. In this one-way street scientific knowledge claimed supremacy over all other forms of knowledge; the task of the scientist was to set out the truth and that of the public to listen, learn and trust. The possibility that such an authoritarian approach was unlikely to convince the better informed and more sceptical publics of the late twentieth century was beyond elite science's understanding.

In Britain, governmental willingness to modify an expertocratic advisory process by opening the process to other voices was accelerated by scandal. BSE/Mad Cow disease, scandals at the Alder Hey and Bristol hospitals, and the panic over the claim that the triple Measles, Mumps and Rubella vaccine could cause autism all intensified public distrust in science. The old claim that science spoke truth to power was increasingly under question. Academic scientists were still trusted more than those working for industry or government, but even here it was draining away.

In addition, the Economic and Social Research Council's programme on the Public Understanding of Science pointed to the limits of solely expert advice on science-related issues and the importance of drawing on the expertise of lay people with direct experience.[22] These studies included Brian Wynne's research on the Cumbrian farming community's understanding of the risks from the radioactive fallout from Chernobyl, Alan Irwin's of a local community's understanding of risk from a nearby chemical factory, and Hilary

Rose's of patients with a genetic disease. Hence the discussion, first in academic, then in policy circles, in both Europe and the UK, of the need to 'engage' the public in the advice-giving process.

While the first three waves of democratising science had come from below (as part and parcel of the attempt to build a science responsible to society and nature alike), the adoption by governments of capitalist countries of 'public engagement' policies – bringing non-experts into the policy advisory process – represented a partial victory for the democratisers. For governments, it was a way of trying to recover trust. In the UK, the public engagement route was eased by an earlier and much acclaimed innovation. To manage the moral crisis posed by the birth of the first IVF baby, and following a lengthy consultation process, the then Conservative government had created a new institution that could and did serve as a model for including 'lay people' in advisory and even regulatory processes. This was the Human Fertilisation and Embryology Authority (HFEA). Faced with more science-related scandals than its European neighbours, the incoming Labour government of 1997 created three new bodies with the same aim of bringing the public into the advisory process – the Agriculture and Environment Biotechnology Commission, the Human Genetics Commission (HGC), and the Food Standards Agency (FSA).

The 2000 report of the House of Lords Select Committee on Science and Technology, *Science and Society*, marked the high point of lay inclusion. The Committee argued that the relationship between science and society was at a critical phase. There was a generalised questioning of and lack of trust in authority, together with doubts about the purposes of science unless its research was seen as 'independent' of government or industry. The Committee pointed out that the prevailing culture of governmental and institutional secrecy in the UK made matters worse, and called for what the think tank Demos referred to as *See-through Science* – where the process of advice or decision making adopted 'a presumption of openness'.[23] For the Research Councils this was near revolutionary; the Medical Research Council (MRC) in particular, as the oldest of them, had been meeting in private for the best part of a century. But while revolutionary for the MRC, just how big a democratic gain would this be?

It was in any case rather easy for government to bypass the new bodies when commercially and ethically sensitive decisions – such as those on human embryonic stem cell research – were required. Ministers retained the right to choose their own advisers independently of the Commissions, a right they exercised in relation to stem cell research, opening the way for Britain to become a leading nation in the field. In the aftermath of GM Nation, a nationwide consultation on GM food which came to the wrong conclusions, the Agriculture and Environment Biotechnology Commission was abolished. The 2010 Coalition government went further, abandoning public engagement by turning the FSA and the HGC into ministerial in-house committees as part of its self-proclaimed 'bonfire of the quangos'. The high tide of public engagement was fast retreating.

SCIENCE IN THE NEW PRECARITY

Even as these experiments in restoring trust in science by opening its decision-making processes to public engagement were under way, successive governments were enthusiastically exchanging the political economy of the welfare state for that of neoliberalism. The rise of the transnational corporations, able to spread production processes across countries, minimising costs and maximising profits, together with the attack on organised labour, began to sap the very foundations on which the welfare state was built. What was prefigured in the crises of the 1970s has accelerated in the decades since, with globalisation, the ideological weakening of collectivism, and the assaults on the achievements of multiculturalism.

With the slow demise of the welfare state, the sense of security of the 99 per cent has eroded and precarity increased, from which, post the crash, only the very rich have made themselves immune. A new era of increasingly dirigiste control has arrived across Europe where research priorities are increasingly set through the EU's multibillion Framework programme. While the political rhetoric claims that the biomedical technosciences will yield immense gains for health, wealth creation is now unabashedly formalised as the chief objective of science and technology policy.

Social precarity has been intensified by the risks of environmental degradation and the attack on civil liberties in the state's response to intensified urban terrorism and social protest, from 9/11 to the train bombings in Madrid and London (directly and predictably stemming from the so-called preventative war against Iraq and the Afghanistan morass). The technologies of these asymmetric wars are being brought back to police the cities of EuroAmerica.[24] What the military theorists describe as low-intensity operations most require is information. Everyone knows about commercial digitalised data gathering and it is impossible not to notice the proliferation of CCTV cameras. But most have become indifferent, while the young openly resist, their hoodies a means of self-defence and their camera-phones a form of counter-surveillance of the police.

Emergent technologies, receiving less publicity, are being built around the fusion of info and biotech, of electronics and bios. First came the biometric markers such as iris recognition, now a familiar airport security device. But despite much research investment in face-recognition technologies, little has yet been delivered. However, the demand by the state for information about its citizens has gone far beyond simply recording their commercial transactions and movements across cities.

Today's high-tech incarnation of Jeremy Bentham's Panopticon observes and records not only our shopping practices, street movements and biometric measures but gathers our biomedical data and above all our DNA. The UK DNA crime data bank, the largest in the world, was curtailed by the European Court of Justice, but the National Health Service (NHS) electronic health records, including DNA information, have been commodified by stealth – tucked into legislation focusing on larger health-care issues. The neurotechnosciences are not far behind, with their promises of being able to read and manipulate mind, memory and intentionality by scanning the brain. Enthusiasts claim that the windows into the activity of the living brain provided by the electroencephalograph (EEG) and functional magnetic resonance imaging (fMRI) can identify potential psychopaths, criminals and terrorists before they have committed a criminal act. Neurosceptics question these claims. Nonetheless the research is increasingly well funded – by the US military for one.

IDENTITY AND THE BIOMEDICAL TECHNOSCIENCES

Within modernity the life sciences have claimed and been given the authority to define 'human nature' (although within this universality biology has a long and deplorable history of inscribing hierarchically ordered difference on human bodies and brains). Ever since Darwin, science as culture has sought to explain our sense of identity. Again and again the molecular biologists leading the sequencing the human genome claimed that the completed genome would constitute human identity. As the historian of science Donna Haraway parodied it, 'genes'R'us'.

The blaze of publicity surrounding both the launching in 1990 and the culmination in 2003 of the Human Genome Project ushered in an era of both genetic determinism and Promethean promise. Gene-talk filled the media, fanned by the brilliant pen of Richard Dawkins with his claim that we are all just lumbering robots driven by our genes. Agency was now attributed to the gene and removed from human beings. The left, anti-racists and feminists fought such biological determinism, but with the likes of President Kennedy's son John maintaining that his drunkenness was due to his having genes for alcoholism, and the geneticist Dean Hamer claiming to have found a genetic marker for homosexuality – immediately if inaccurately christened the gay gene – the geneticisation of culture had indeed gone a long way. Genes for everything from criminality to compulsive shopping soon followed. By the end of the first decade of this century, behaviour geneticists have held genes to be responsible for everything from sexual orientation, female coyness and male violence to voting intentions, respect for royalty, Battersea road art and the inevitability of a neoliberal economy.

Such claims, amplified by a largely uncritical media, have entered into popular culture. Everyday gene-talk, with its genes for this and genes for that, has powered up, and now that everyone knows that genes are made of strings of DNA the metaphor has flown into orbit. It is to be found in the design of BMW motors, the core values of Cameron's Conservatives, and most recently those of Putin's Russians: 'We are a victorious people – it's in our genes, in our genetic code, passed down from generation to generation', Putin declared at his presidential inauguration in 2012. Indeed it is hard to find a politician or media pundit who fails to invoke the metaphor in almost any conceivable context in which immutable values are being invoked.

The neurosciences have not been left behind; their claims to explain selfhood, love and consciousness as located in specific brain regions – a sort of internal phrenology – have been articulated in a string of popular books. Joseph Ledoux writes of *The Synaptic Self*, Antonio Damasio of *Descartes' Error*.[25] Semir Zeki, professor of neuroaesthetics, claims that romantic love is the product of neural activity in the putamen and insula, while Francis Crick in *The Astonishing Hypothesis* argues that our very intentionality and agency as humans is an illusion and in 'reality' we are nothing but a bunch of neurons, our consciousness closeted in the claustrum and free will in the anterior cingulate sulcus.[26] Jean-Pierre Changeux writes of *Neuronal Man*.[27] Given current sensibilities about the directive power of language, 'neuronal man' jars, but it is autism researcher Simon Baron-Cohen who has been the most recent to insist on what he calls *The Essential Difference*: that there is both a neuronal woman and a neuronal man, their differences fixed by a hormonal surge to the foetal brain.[28] Such essentialist claims have been strongly criticised not just for their methodology but for their underlying premises; most recently by two of a new generation of feminist neuroscientists – Cordelia Fine in *Delusions of Gender* and Rebecca Jordan-Young, in *Brainstorm*.[29]

More than half a century ago Jean-Paul Sartre rejected such claims: 'there is no human nature ... man is nothing else but what he makes of himself'. In *The Second Sex*, Simone De Beauvoir insists that 'one is not born, but rather becomes, a woman'. Hannah Arendt spoke of the 'human condition', though today many would prefer to speak of human conditions in the plural. In the *Eighteenth Brumaire*, Marx wrote: 'Men make their own history but they do not make it as they please; they do not make it under self-selected circumstances, but under circumstances existing already.' This needs extending, however; those 'circumstances existing already', by which both women's and men's actions are constrained, include both the human history and the current social conditions to which Marx refers, but also, as his contemporary Darwin insisted, the history of human biology itself. Both these giants of nineteenth-century social and biological theory (setting aside the anomaly of Marx's commitment to stages of historical progress and some of Darwin's progressivist hopes) were radical indeterminists. We share that indeterminacy; humans can make their own history, but they do so in circumstances which include both their embodied social existence and their socially embedded biological existence.

The dominant voices of the molecularised life sciences of the twenty-first century are unhappy with such complexity and retreat to a simply biological narrative. Their discourses are at once essentialist and Promethean; they see human nature as fixed, while at the same time offering to transform human life through the real and imagined power of the biotechnosciences. These are the issues we will examine critically in the chapters that follow, beginning with the Human Genome Project.

1

From Little Genetics to Big Genomics

In 1992, at the start of the surprisingly short decade-long march towards the sequencing of the human genome, one of its key initiators, geneticist Walter Gilbert, pulled a glittering CD out of his pocket and held it up to his audience announcing: 'soon I will be able to say "here is a human being; it's me".'[1] Gilbert's brilliant piece of theatre was echoed by other leading molecular biologists in their campaign to win public support and enthusiasm for the Human Genome Project (HGP), the ambitious international effort to sequence the three billion nucleotides that constitute the human genome, at a cost then estimated at $3 billion – a dollar a nucleotide. It seemed not to matter how often molecular biologists employed the same theatrical device, whether in California or at London's Institute for Contemporary Arts; holding up a CD to their spellbound audiences and saying this is human life itself was a brilliantly chosen trope. The CD, so familiar to the audience of a 1990s high-tech society, was recruited to symbolise the merger of molecularisation and digitalisation of biomedical research heralded by the HGP. Human genomics simultaneously offered a new definition of human nature, and new Promethean powers to repair and even re-engineer that nature.

THE DOUBLE HELIX: ACHIEVEMENT AND ERASURE

It had been a long march since 1953, when the Cambridge based biologists Francis Crick and Jim Watson described the now famous double helix structure of DNA, composed of vast linked chains of four small molecules, the nucleotides – or bases – adenine, cytosine, guanine and thymidine (A, C, G and T). At King's College London, the biophysicist Maurice Wilkins and the X-ray crystallographer Rosalind Franklin were also working on the structure of DNA. Thanks to their head of department's bad management they had each separately been given the same problem to work on. Wilkins assumed he was in charge, and passed on Franklin's critical data to the Cambridge pair. She was unaware that her images of DNA crystals had been 'shared' – or more bluntly, stolen – without her knowledge or consent.

With the last sentence of Crick and Watson's path-breaking *Nature* paper, 'It has not escaped our notice that the [specific] structure we have postulated immediately suggests a possible copying mechanism for the genetic material',[2] DNA's career was launched. While the significance for the life sciences was immediately recognised by the international scientific community, there was little public interest. The media did not pick up on its significance, and it was decades before the life scientists learned to practice megaphone science, piling hyperbole on hyperbole, heralding the birth of the HGP.

It was Watson's *The Double Helix*, published fifteen years later, after both Franklin's death and the three men's shared Nobel Prize in 1962, which put DNA into the public arena. When the three gave their acceptance speeches at the Nobel Ceremony none spoke of their late colleague's contribution. Watson's book unblushingly revealed the arrogance and ambition of the Cambridge pair, the rivalries, personal jockeying for position, unauthorised acquisition of data and his own total inability to behave appropriately towards Franklin as a senior scientist whose crucial X-ray skills had made the model possible. Many scientists saw the book as bringing science into disrepute, thereby revealing more about their longing for a lost golden age than their knowledge of the history of science. They had erased the memory of so many savage disputes in the past – notably Newton's battle with Leibniz to secure the prior claim to having invented calculus. As President of the Royal Society, Newton appointed a committee to adjudicate on the matter and drafted its predictable conclusions himself. The charge that Watson had smirched the good name of science was unconvincing, as conference gossip among researchers enjoyably recounts such conflicts, why the results from X's lab should be taken with a pinch of salt or why scientist Y has reason to believe that an old enemy Z was the referee who had delayed their paper to secure priority over Y. From this perspective Watson let the side down by informing the public of what only scientists were supposed to know. In his autobiography Wilkins notes that, at the time, he and Crick had talked of suing Watson,[3] who became known, not entirely to his comfort, as Honest Jim. Meanwhile,his account of the inner workings of science – as slightly less than 'pure' – sold like hot cakes.

While the injury to Franklin is now publicly recognised and has been symbolically corrected – one of the new King's buildings in London is named the Franklin Wilkins Building – what is less well known is how the injustice was exposed. The writer Anne Sayre, married to X-ray crystallographer Donald Sayre, was a close friend of Franklin. Both she and her husband and indeed many crystallographer colleagues were profoundly angered by Watson's unjust and insulting treatment of Franklin. Drawing on this network, Sayre published a biography of Franklin in 1975.[4] The story was seized upon by the nascent feminist movement as providing unambiguous evidence that Franklin's male colleagues had systematically diminished her contribution, and that Wilkins had de facto stolen her X-ray photographs. Wilkins was sufficiently disturbed by this claim that he asked one of the London Women and Science group, whether she thought he had in fact done so. Watson characterised Franklin as not only difficult, but also incapable of recognising the theoretical significance of her own images. Only the theoretically minded (i.e. men like himself and Crick) were up to the job. To this Watson added an unpleasant sexist commentary on her appearance. It was left to a second biographer, Brenda Maddox, to point out that Franklin in fact dressed very elegantly, having developed an interest in fashion after working in a laboratory in Paris.[5] But her dress aesthetic was manifestly outside Watson's capacity to recognise, let alone appreciate. Watson himself poured petrol onto the flames by observing that 'the best place for a feminist is someone else's lab', although ironically his unwitting exposure of the institutionalised sexism within science nourished the growing feminist criticism of science.

LITTLE SCIENCE, BIG SCIENCE

In the 1950s Watson's declared ambition was a Nobel Prize, not yet patents and profits. His research was university-based, relied on one powerful technology – X-ray diffraction – and cardboard or tin cut-out models of the four bases, A, C, G and T. Genetics was then, to use Derek de Solla Price's term, a Little Science.[6] Little Science had a low ratio of capital to labour. Laboratory work was hands-on, craftwork, from synthesising reagents to building equipment, from pipetting solutions to washing the glassware at the end of the experiment. Papers were frequently single-authored, or at most bore two or three names.

Over the subsequent decades, genetics and molecular biology have been irreversibly transformed into Big Science. For genomics – the analysis of entire genomes – to arrive required the fusion of human genetics and molecular biology, a fusion only made possible by the new informatics. With this new configuration the capital–labour ratio was reversed. The manufacture of equipment and the production of reagents has been outsourced to specialist firms, and laboratory practice has become automated. Today's sequencing laboratory consists of a huge bleak space with identical machines about the size of a domestic microwave spaced out along the benches. A gowned technician silently tends the machines, while the researchers sit in their offices reading the printouts. Nothing could be further from the bustle of both scientists and technicians competing for space in the busy labs of the 1950s.

The papers coming out of the big sequencing labs can have several hundred authors – even entire institutes. Knowledge has become intellectual property and must be patented. The research investment is too great for it to be left to chance whether the media will or will not cover the story. Press releases are now carefully crafted by the lead researchers together with the university press office. Meanwhile the journal publishing a paper also works to amplify its impact and thereby reinforce its own status as a prestige journal, first by online prepublication of the paper and then by providing a summary written by a science journalist to explain the paper's importance in its field and its possible contribution to health care and wealth creation. These new practices of science communication position the research most favourably in a market where knowledge has become intellectual property. Although the major newspapers all have science correspondents, 'churning' – reproducing a press release – has become a common media practice. When this happens the science writers of the fourth estate fail in their critical role and become part of megaphone science.

Meanwhile research is on the move both physically and financially, away from the universities towards spin-off companies and beyond. Big Pharma has got bigger as mergers first hyphenate and then delete once famous names. Burroughs Wellcome became Wellcome became Glaxo-Wellcome became the conglomerate GlaxoSmithKline, leaving the name Wellcome only to the giant London-based charitable trust. This revolution has been fuelled by the hopes of many that biotechnoscience will generate genetic diagnoses that will lead smoothly into clinical interventions, from gene therapy to new drugs. All the known problems of translation, of getting biomedical research innovation from bench to bedside, have been smothered in a sea of hype. In the process, leading biologists and geneticists have become entrepreneurs, interested as much in the value of their stocks and shares as in the smooth and successful running of their laboratories.

This shift in the entire production system of the life sciences with the entry of entirely new stakeholders with different cultural values is exemplified by Watson's own career path. In the 1990s, as the Director of one of the powerhouses of US molecular biology, Cold Spring Harbor, he was told by one the researchers in the lab, Tim Tully, of a promising new discovery about a molecule that might enhance memory. His scientific excitement, Tully reported, was equalled only by his delight at the prospect of patents and 'shed-loads' of money. (Watson's often quoted and typically caustic interventions in the genomics narrative serve as both benchmarks and catalysts.) Tully, very much one of the new bio-entrepreneurs, responded by spinning out a private company and then leaving Cold Spring to work for it. Fuelled by high-profile praise in the media, Tully became – briefly – a celebrity scientist. But as the hopes for his molecule faded, the firm, like so many start-ups begun with high hopes and readily available venture capital, crashed. The fact that his project later re-emerged phoenix-like from its ashes, refinanced and relocated to California, is indicative of the extraordinary ups and downs of biotechnoscience within globalisation.

The HGP would have been inconceivable without this transformation, as the painstaking stepwise analyses that had characterised the early days were swept aside by robotic instrumentation and giant computers requiring major financial support. Routine work – of the sort that Watson once derided as capable of being done by a team of monkeys, and that would have taken thousands of skilled technicians decades to finish – could now be completed by machines in weeks or days. Even the superfast technology of the 1990s and the early years of this century now seems a remote memory.

A crucial step in this transformation of science was a decision made in 1980 by the US Supreme Court. Ananda Chakrabarty, a microbiologist working for General Electric, had genetically engineered bacteria able to break down oil – thus with the potential to clean up oil spills – and had sought to patent it. The US patent office (in the name of its officer, Diamond) had earlier rejected the claim on the grounds that a living organism could not be patented, but was overruled by the Supreme Court. (Two decades later a legal judgment in a lower court in the US effectively reversed the Diamond v. Chakrabarty ruling, but it is unlikely that this will conclude the patenting fight, as too much money is at stake.)

In the same year, the US Congress passed the Bayh–Dole Act granting universities intellectual property rights over inventions and products arising from federally funded research. These two decisions opened the floodgates. Now it seemed anything could be patented – bugs, constructed animals such as the Oncomouse (engineered at Harvard, designed for cancer research), and even strands of DNA. The first patent for a gene, that for insulin from genetically engineered yeast, was granted to the biotech company Genentech in the same year. In the context of a near pandemic of obesity and its consequence, diabetes, the importance of this patent is evident. Bayh–Dole and Diamond–Chakrabarty are the twin markers of the transformation of a once disinterested science into one that is very much interested. Today, for the leading sequencers of the HGP and their successors, prestige, with the Nobel Prize still the ultimate symbolic achievement, remains. But money has joined prestige and become entangled with the control of intellectual property, patents, and access to hugely profitable stakes in instrumentation, biotech and pharmaceutical companies. This new hybrid production system of science is radically different from that of the past.

TO MAP OR TO SEQUENCE?

DNA has two distinct functions in the life of the individual and of the species. As Watson and Crick had immediately realised, its double helical structure means that, when unwound, each strand of the helix serves as a template on which the other can be built – resulting in two identical molecules of DNA. This makes possible the transfer of genetic information, coded for in the strands of DNA, from one generation to the next during reproduction. The second function is its central role in the cellular economy. The code written in unique sequences of As, Cs, Gs and Ts is read by the cell via an intermediate, RNA (a molecule similar to DNA), to direct the synthesis of proteins. At the time of the Watson–Crick paper it was assumed that one gene coded for one specific protein. Crick called this the Central Dogma of molecular genetics: 'DNA makes RNA makes protein makes us'. He described DNA as an informational macromolecule, and saw the flow from DNA to protein as a one-way transfer of information. Again, it was Crick who put it most succinctly: 'Once information has got into the protein it can't get out again.' Knowing the DNA sequences, the molecular biologists believed, would provide the key not merely to understanding genetic mechanisms but also to intervening in them, manipulating both cell metabolism and heredity. But sequencing at the time, in the 1970s, presented a formidable technical problem. There was, however, a more promising approach: mapping.

In humans the DNA that comprises the genes is located within twenty-three pairs of chromosomes, present in the nucleus of every cell in the body. One of each pair is inherited from each parent. The chromosomes are individually recognisable under the microscope by their characteristic shape. One pair of chromosomes differs between males and females; females have two X chromosomes, males one X and one Y; the other pairs are numbered one to twenty-two. Long before they were known to be made of DNA, techniques for mapping where the genes lie along the chromosomes had been pioneered in what was for many years the geneticists' favourite organism, the fruit fly. Mapping and identifying the genes in human chromosomes was technically much harder, but the clinical interest in identifying disease-associated genes made it the research priority of the 1970s, and a substantial number of disease-related genes had been mapped to particular chromosomal regions.

In the decades following his formulation of the Central Dogma, the elegant clarity of Crick's slogans became distinctly fuzzy. Not merely were the information flows less unidirectional than he had imagined, but it had become clear that genes coding for proteins only constitute a tiny proportion – less than 2 per cent – of the DNA in the human genome. The remaining 98 per cent of the DNA in the genome was disparagingly referred to by molecular biologists as 'junk', as it was thought to have little or no biological function. So why sequence the entire genome, as the HGP proposed, if much of the sequence would be 'junk'? The term, however, proved to be seriously misleading, with painful consequences for the hopes of the HGP as 'the book of life', as we explain in the next chapter. But at the time when the HGP was being contemplated, it was the genes, not the junk, which offered clinical hope and raised the possibility of future patents.

Some 7,000 different diseases – some extremely rare, affecting only a few families worldwide, and others common, like cystic fibrosis (CF), which affects one in 3,000 newborns among Europeans, or graver still, sickle cell, affecting two in a hundred live births among Africans – are inherited in a Mendelian manner. That is, they are transmitted through a single mutation affecting one of the many genes necessary for the biochemical pathways that are disrupted in the disease. For some diseases the mutant gene is dominant – that is, one copy (called an allele) is sufficient to cause the disease; in others the gene is recessive and two copies, one inherited from each parent, are required. A person is said to be homozygous if they possess two identical copies of the gene, heterozygous if they have only the one. In addition, some conditions, like colour blindness, and some diseases, like Duchenne muscular dystrophy, are transmitted by a recessive gene carried in XX women, but only manifest themselves in XY men.

Labs around the world were hunting such disease-associated genes. The goal was straightforward: once the gene has been found – or at least mapped – it should be possible to produce a (patentable) test for diagnosis or screening. This can be immensely profitable for biotech companies. The gene *BRCA1*, which predisposes towards breast cancer, was identified in 1990 and cloned in 1994 by the US-based Myriad Genetics, which patented it for the international market, and was then for many years able to make large profits by charging $3,000 per test. The French challenged the validity of the DNA sequence on which Myriad had based their test, and simply bypassed the patent. French DNA sequences, they said, were different. Other European countries accepted the patent's validity and paid Myriad's charges. In 2010 the patent was also successfully challenged by the American Civil Liberties Union. Given the amount of money at stake, the legal battle, like that over Diamond–Chakrabarty, is likely to continue. In the twenty-first century, the ethics of patenting life are being questioned.

Beyond such tests lay the prospect of gene therapy – replacing the mutant gene with a properly functioning one – and it was on this that the future of the burgeoning biotech industry seemed initially to rest. However, diagnosis proved to be much easier than therapy. The technologies for gene replacement in an adult, or even a young child (through somatic gene therapy, which only affects the body of the treated patient and is not passed on to any children), have proved technically difficult, hazardous to health and ethically problematic.

Nonetheless, to the majority of human geneticists and their backers in the 1970s and '80s, gene mapping seemed the way forward. Ahead of the field was a French group, CEPH (the Centre for the Study of Human Polymorphisms), directed by geneticist Daniel Cohen, and supported by both government grants and an extraordinarily successful telethon appeal to the public from the French Muscular Dystrophy Association. The success led to the establishment of Généthon, a non-profit organisation based near Paris, which by the early 1990s had produced a map of Chromosome 21 far ahead of anyone else.

It was not just that mapping seemed a better strategy than sequencing – it was likely to prove a great deal easier. The techniques for reading the order in which the As, Cs, Gs and Ts of the code occurred were manual and very labour intensive. They had been developed by Frederick Sanger in the Medical Research Council Laboratory of Molecular Biology (LMB) in Cambridge, where Crick and Watson also worked. Sanger had already received a Nobel Prize in 1958 for his sequencing of a protein, insulin, and from the 1960s on he worked out the methods to read the nucleotide sequences first in RNA and then DNA. Descendents of his analytical methods form the basis of all the automated modern sequencing methods. Essentially, they depend on first preparing a pure sample of the DNA, then breaking it down into short chains, which can be separated by size, and sequencing each individual chain. Finally the short sequences have to be fitted together by means of computer algorithms to provide a readout of the order of the bases in the full chain of the original DNA. It took Sanger some fifteen years to perfect the method and to publish the sequence of the 48,500 bases in a bacteriophage (a virus that infects bacteria). In 1980, he won a second Nobel for the work, this time shared with the US molecular biologists Paul Berg and Walter Gilbert, both of whom were to play leading roles in the transformation of genetics into the Big Science it is today.

Sanger's breakthrough opened the way for a number of small-scale sequencing projects. Perhaps the most significant in view of the subsequent history of the HGP was that initiated by Sanger's colleague at LMB, Sydney Brenner. An early collaborator with Crick, in the mid-1960s Brenner had begun a project that seemed then as unpromising as the HGP did twenty years later. He was convinced that the next major biological breakthrough after DNA would stem from choosing a simple multicellular organism and discovering all one could about its anatomy, behaviour and genetics. He chose a tiny nematode worm, *C. elegans*, about a millimetre long, and with only 959 cells in its body. Unimpressed biologists described Brenner's project as the most boring serial electron microscopy project in the history of science, and the multitude of PhD students and post-docs he gathered round him as little more than assembly workers in a Fordist factory. Fortunately for Brenner however, the MRC shared his vision and funded the work. By the 1980s he had set one of his younger colleagues, John Sulston, the task of sequencing the 100 million bases of the worm's DNA. Sulston's skill and tenacity, plus the prestige of the LMB itself, meant that his lab was to become the focal point of the British share of the HGP, heavily funded by the Wellcome Trust. (Not for nothing did religious writer Andrew Brown title his account of Sulston's work *In the Beginning Was the Worm.*[7]) Now the largest charitable funder of biomedical science in the world, Wellcome's role as a major actor in the HGP was not a one-off intervention in genomics; it has continued to fund and shape subsequent large-scale developments, not least the UK Biobank (see Chapter 6). For British life sciences, the Wellcome Trust, in setting the agenda and forcing the government and MRC to follow its line, currently plays a similar role to that of the Rockefeller Foundation, which shaped the development of biochemistry in the 1930s (as we discuss in the next chapter).

Since the human genome is some 6,000 times as long as that of Sanger's bacteriophage, and thirty times greater than that of Brenner's worm, once again many biologists regarded sequencing its entire length as being of doubtful scientific value. So while sequencing continued, it had a lower priority than mapping. In an international though still informal collaboration, several academic labs in Europe, Japan and the US partitioned out the work of decoding specific human chromosomes. During the late 1980s, under the leadership first of James Wyngaarden from the US NIH (National Institutes of Health) and later of geneticist Walter Bodmer, director of the London-based Imperial Cancer Research Fund (now Cancer Research UK), an international body, HUGO, the Human Genome Organisation, was established to coordinate these efforts. Only with the launch of the HGP did the sequencing project come to dominate.

THE IMAGINARY OF THE HGP

The story of the origins and ultimate success of the HGP is one in which unparalleled hype, institutional rivalries, the promise of profit, technological brilliance, and scientific hubris intersect. One of the earliest molecular biologists to propose sequencing the entire human genome was Renato Dulbecco, Nobelist and President of the Salk Institute in California (named after Jonas Salk, creator of the polio vaccine). Dulbecco was not seeking a grand project for the sake of prestige, but through his involvement in President Nixon's War on Cancer. His claim that sequencing genes offered the most realistic possibility of effective cancer therapies caught the imagination. Sequencing became a hot topic among the international molecular biology community.

The first attempt to raise the funds for the project was made by Robert Sinsheimer, a molecular geneticist who had won acclaim for his pioneering synthesis of a functional strand of DNA ('the closest anyone has yet come to creating life in the laboratory' was how the press releases described his achievement) who had been one of the early movers in the burgeoning biotech industry. In 1985, as Chancellor of Santa Cruz, one of the smaller Californian universities, he sought a major research project to enhance its status. The project he eventually hit on was the genome, and he tried to persuade a wealthy donor to fund the sequencing. The proposal foundered – the funding required was more than any individual donor could meet, and institutional bickering among the component universities in the Californian state system finally killed it.

But Sinsheimer's project, despite its failure, inspired Walter Gilbert to try another approach. Gilbert was the most eminent among the new breed of geneticist-entrepreneurs. His career had moved seamlessly between Harvard, his company Biogen, and then – when Biogen crashed – back to Harvard. Fired up by the Santa Cruz proposal, it was Gilbert who first described the decoding of human DNA as biology's 'Holy Grail', just one of a plethora of Christian metaphors in which the sequencers were to enshroud their quest. Gilbert announced in 1986 that he was going to form a purpose-built company – Genome Corporation – to do the job. He too failed to raise the huge $3 billion required, and Genome Corp. sank without trace.

Gilbert's efforts to commercialise genomic research have been frequently erased in the origin narrative of the HGP. Occasionally geneticists suffer mild amnesia concerning this initial bid. It seems they prefer an account that begins with the non-commercially tainted public sector. But it was only after these preliminary attempts had failed that two major US federal funders entered the scene. One was predictable, the giant NIH – the other, quite startling in its apparent distance from the field, was the Department of Energy (DoE). The DoE's origins can be traced back to the 1940s, with the development of atomic weaponry and later the oversight of civilian nuclear power. The DoE's remit included a continuing study of the survivors of the atomic bombs that devastated Hiroshima and Nagasaki, especially the medical and genetic consequences of their exposure to radiation for their descendents. By the 1980s the study of survivors was not yielding clear enough outcomes to satisfy policy makers; in consequence, the incoming director of DoE's biomedical research division saw the need for a new big genetics project to ensure the division's future. One eminent if less than tactful geneticist sardonically described it as 'a project for redundant bomb-makers'.

Securing and maintaining public and political support – thence the funding – for this first Big Science project in the history of biology was a priority. Thus throughout the life of the HGP the scientists' rhetoric continuously fused hope and hype, recklessly comparing their project to 'the discovery of fire' or the 'invention of the wheel'. Scientists no longer needed the media to inflate their claims, nor did they await laboratory testing, let alone the judgement of history. Certainly the 1989 editorial in *Science* by the biochemist Daniel Koshland celebrating the launch of the HGP showed little restraint:

> The benefits to science of the genome project are clear. Illnesses such as manic depression, Alzheimer's, schizophrenia, and heart disease are probably all mutagenic and even more difficult to unravel than cystic fibrosis. Yet these diseases are at the root of many current societal problems. The costs of mental illness, the difficult civil liberties problems they cause, the pain to the individual, all cry out for an early solution that involves prevention, not caretaking. To continue the current warehousing or neglect of these people, many of whom are in the ranks of the homeless, is the equivalent of providing iron lungs to polio victims at the expense of working on a vaccine.[8]

Koshland's vision goes far further than merely relieving the sequencers from the small problem of successfully moving biomedical research from bench to bedside. In his lyrical affirmation of the HGP he sees beyond the mere reduction of disease to the solution even of homelessness. But in his explanation of how this may be achieved through 'prevention' his editorial opens the door to nothing else than the revitalised eugenics we will discuss in Chapter 4.

Robert Cook-Deegan, writing from the vantage point of his term in the US Office of Technology Assessment, has chronicled in meticulous detail the departmental infighting over the DoE's sequencing proposals, which eventually resulted in agreement on a joint project, funded between the DoE and the NIH.[9] An overall leader for the project was required and, as Cook-Deegan recounts, Watson stepped shyly into the limelight – as indeed had been predicted by John Sulston, who saw Watson as seeking power over the direction of the entire field of genetics. At the press conference announcing his appointment as director of the HGP, Watson was asked how he proposed to address the social implications of the project. As there had been no prior consideration of this issue, Watson, thinking on his feet, announced that 5 per cent of the NIH-HGP budget would be set aside for the study of its Ethical Legal and Social Implications (ELSI). In the years that followed, the huge funds from ELSI were to influence the development of the humanities and the social sciences, much as the HGP was to influence the development of the life sciences.

However, Watson's reign lasted barely four years, when a very public disagreement with Bernardine Healy, the NIH director, and a conflict of interest over his wife's shareholdings led to his resignation. His successor was Francis Collins, prominent both as a geneticist and as an openly committed Christian (not a common combination) who describes in his autobiography how he spent time in prayer before accepting the task.[10] Taking over in 1993, Collins reorganised the project. The genome was now to be sequenced over a decade, through a collaborative international (in practice mainly US–UK) effort. The major centre outside the US was Sulston's group, based at a dedicated lab at Cambridge, the Sanger Centre (now renamed the Sanger Institute), funded by Wellcome and the MRC.

The intention of the HGP was to produce a consensus sequence, derived from the DNA of a small number of anonymous individuals, so as to smooth out individual differences as far as possible. As sections of the DNA were completed, the data was to be deposited in GenBank, a publicly accessible free database held by NIH's National Center for Biotechnology Information. Despite Sulston's genuine belief that this was a 'gift to humanity', for the funders the policy was less about high-minded idealism than the worry that premature patenting would slow down innovation, hindering the drug industry's development of new drugs. Thus their problem was not with patenting per se, but with premature patenting. In this way the vast public subsidy would yield valuable results, profitable for Big Pharma and hence for US and UK plc.

ENTER VENTER

However, the carefully negotiated plan for the public funding and international division of labour was soon to be rudely shattered by the appearance on the scene of a young, ferociously ambitious and highly talented biochemist-turned molecular biologist, J. Craig Venter, at that stage working at the NIH. Venter rejected the slow methodical approach offered by the traditional sequencers. He began by proposing a quick way to identify the genes embedded within the genome by an ingenious technique he called expressed sequence tags (ESTs), and proposed publishing the method. As the tags had been artificially synthesised rather than merely discovered in nature, NIH, his employer, wanted to seek patents for them. Venter objected, preferring the method to remain in the public domain, but NIH went ahead anyhow, filing patents in 1991 on 347 ESTs his team had discovered.

Venter then proposed a still faster way of sequencing (shotgun sequencing), which involved breaking the DNA into much shorter lengths than had Sanger's original procedure and using massive computing power to reassemble the sequence. Sulston and others derided the method, claiming it would result in a sequence full of errors. Watson, however, was attracted by Venter's shotgun approach, and encouraged him to put in large grant proposals. But on four successive occasions, so Venter says in his autobiography, Watson was unable or unwilling to convince his committee to fund them.

Was it bad luck or foul play? After all, this wasn't the first time the two had clashed. Early in Venter's career, Watson had commented on one of his research findings: 'That explains everything; You're a biochemist.' Venter tells us that initially he saw it as a compliment, not being aware of the long battle for intellectual supremacy that had taken place between biochemists and molecular biologists. Biochemistry was the older discipline, and biochemists objected to the newcomers from physics and engineering, like Watson and Crick, muscling in on their territory, especially when they began to solve problems that biochemists could not. When Watson and

Crick published their DNA paper in 1953, Erwin Chargaff, a leading DNA researcher, memorably accused them of 'practicing biochemistry without a licence'. By the 1970s the tables had been turned. It was molecular biology that was now the glamorous science; biochemistry, as the mere chemistry of life, was becoming old hat. (As we discuss in the next chapter, it took another couple of decades for yet another turn of the tables to occur, as molecular biologists rediscovered biochemistry, rebranding it as 'proteomics'.)

Frustrated, short of the funds needed to meet his ambitious goals, and fed up with the non-trivial task of writing large-scale research proposals, Venter turned to the private sector. The first move, in 1992, was to set up a new lab, TIGR (The Institute for Genetic Research), a not-for-profit organisation funded through the company Human Genome Sciences (HGS). HGS, a drug discovery and marketing company, was the creation of William Haseltine, a former AIDS researcher turned major entrepreneur. (In his younger days Haseltine had been an anti-Vietnam war radical, exposing the US military's new automated weapons systems in a booklet and slide-show called *The Electronic Battlefield*. Venter had been a young paramedic stationed in Vietnam, both shocked and toughened by the battlefield wounds he was called upon to treat. It was this experience that took him back to university and thence to research.) HGS would have first call on TIGR's data, which would only be published with a six-month delay to give Haseltine time to patent anything promising. The stage was thus set for a race between the public and private projects, a race marked by increasingly vituperative exchanges between the two sides. Venter compared Sulston and his team to mediaeval monks, working with quill pens. Sulston and Collins saw Venter as an arrogant cowboy. Relations were not improved when, in 1995, ahead of other academic researchers and in a blaze of publicity, TIGR published the first complete sequence of a simple bacterium, *Haemophilus influenzae*, a full 2 million bases long.

By 1998, however – by which time the public effort had only managed to generate some 4 per cent of the human sequence – Venter saw an even more tempting opportunity. Leaving TIGR in the charge of the molecular biologist Clare Fraser, to whom he was then married, he became President of a new company, Celera Genomics, set up with a $950 million capitalisation as a deliberate private rival to the public HGP. Celera proposed to complete the sequence using shotgun sequencing at a tenth of the cost – that is, for about $300 million – and in half the time of the public programme. Celera's sequence would also be derived from the DNA of five anonymous individuals, though it soon became an open secret that much of the DNA was Venter's own. The assumption was that the human sequence should not be regarded as a reference against which individual variations might be matched but rather as a consensus representative for all of humanity. Once again, in publicising his choice of DNA donors, Venter wrong-footed his public competitors and showed his PR skills.

To achieve his goal he needed speed, and speed was what Celera provided, as it was a subsidiary of a larger group of scientific instrument manufacturers, PE (later renamed Applera). Working at another PE subsidiary company, Applied Biosystems (ABI), was a former colleague of Venter's, Mike Hunkapiller, who had modified Sanger's methods to greatly speed up the sequencing. ABI built the automated sequencing machines, called sequenators, and Celera recruited computer wizards to create the programs that could stitch the short sequences together as fast as Hunkapiller's machines could generate them.

Rattled – and angered – by Venter's move, the public sequencers were forced to up their game. Without the massive financial intervention of Wellcome, sometimes spoken of as the ten-thousand-pound gorilla in the genomics room, it is far from clear that the public project would have been driven through. Eschewing Venter's shotgun methods, Collins, Sulston and their team nonetheless also turned to Hunkapiller's sequenators, buying them in bulk, recruiting their own programmers and releasing their data into the public domain practically as soon as the printouts spewed from the machines. However, their releases also aided Venter, since he could incorporate the public data into his own sequences, while denying the public sequencers access to his own, proving that secrecy can be as useful to a scientist-entrepreneur as patenting.

In a PR attempt to gloss over the disputes, which were attracting increasing and unwanted press commentary, the race was declared a tie in 2000, when the two 'first drafts' of the genome were published – the public one in *Nature*, the private one in *Science*. President Bill Clinton, flanked by Collins and Venter, and Prime Minister Tony Blair, speaking over a video link and flanked by no one, held a joint press conference announcing this new gift to humanity. Sulston, to his irritation, was neither present nor referred to by Blair. As scripted by Collins, Clinton referred to the sequence as 'the language in which God created life' – a sentiment with which one assumes Blair concurred. Neither noted the closing words of the *Nature* paper announcing the sequence, which echoed those of Watson and Crick in 1953: 'It has not escaped our notice that the more we learn about the human genome, the more there is to explore.'[11] Once the transatlantic diplomacy and jockeying involved in this theatrical (and somewhat premature) press conference was over, the public consortium got back to work, declaring the 'finished' sequence complete in 2003 and formally winding up the HGP.

The jockeying was much more to Venter's taste than Sulston's, or for that matter Collins – as the account in his autobiography gleefully describes.[12] But what Venter also makes clear are his truly Promethean ambitions and his flair for showmanship – integral to securing the resources to realise them. In 2007 he became the first individual to have his entire genome sequenced and placed in the public domain. The second, predictably, was Jim Watson. Collins also discusses some of his own genome in *The Language of Life*.[13] It is perhaps understandable that the custodians of the old disinterested approach to science, the Nobel committee, chose to award the 2002 prize not to Venter but to Sulston, together with Brenner and Robert Horvitz – and then not for the human genome but for the worm. In sharp contrast to Venter, who has continued to claim ever broader reaches of biology in his attempts to synthesise artificial life, Sulston on retirement withdrew from the laboratory to chair the Institute for Science, Ethics and Innovation in Manchester, and to campaign against gene patenting. His account of the race for the human genome could hardly be more different in style from

Venter's.[14] But the dominant voice of the technosciences of life in the twenty-first century is that of Venter not Sulston.

THE HUMAN GENOME DIVERSITY PROJECT

The revelation of the complexity of the human genome came as a surprise, at least in part due to the sequencers' failure to recognise the significance of the fact that the genome is the product of 3.5 billion years of evolutionary history. Within a decade of the Clinton–Blair press conference, several other mammalian genomes had been sequenced. The almost embarrassing closeness of the human sequence to that of our near evolutionary neighbours, the chimpanzees, as against the obvious differences between any human and any chimpanzee, only highlighted the fact that, despite Gilbert's claim, whatever was meant by being human, it was not simply embedded in a DNA code that could be burned onto a CD. Much, much more was required.

The HGP's scientific assumption was that to study the human genome, as against say that of the gorilla, researchers could combine DNA samples from relatively few people. Neither the individuals' cultural identities nor their obvious physical differences raised theoretical problems in this pursuit of the representative human genome. Who they were didn't matter. However, because human beings are unique (even identical – monozygotic – twins are not quite identical), so are their individual genomes. This gave the sequencers the technical but soluble problem of dealing with the slight variations particularly in the non-coding regions of the gene, primarily changes in single letters of the DNA alphabet – called single nucleotide polymorphisms or SNPs (pronounced snips). Venter, ever the skilled publicist both for Celera and himself, made it clear that while he signed up entirely to this core assumption of the Human Genome Project he also played into, indeed celebrated, the diversity of the US population by using the DNA from five ethnically and gender diverse people. The cover of the issue of *Science* announcing his sequence was graced with idealised portraits of his five symbols of diversity, while *Nature*, the home for the public sequence, was left with an abstract representation of thousands of faces, celebrating universality. Venter had even more fun when he later explained how much of his own genome had gone into the sequencing work. By contrast the public genome sequencers had invisible donors. They thus missed out on Venter's skilful celebration of multiculturalism.

The Human Genome Diversity Project (HGDP), which focused on human evolution and variation over time, was an attempt to go beyond the 'representative' genome that the HGP offered. Although proposed as a parallel project to, and augmentation of, the HGP, and turning initially to the same funders for support, the HGDP was a separate project. The prime mover was the Italian-American human population geneticist Luigi Luca Cavalli-Sforza, who, along with Walter Bodmer, had been a highly visible figure in the scientific opposition against the recrudescence of scientific racism in the 1970s (discussed in Chapter 4). However even during that explosion of social movements, it did not follow that scientists who criticised racist science as bad science would necessarily have recognised the civil rights movement's claims for human agency. Science speaks from above; social movements speak from below. In the case of the HGDP this failure of comprehension was to plunge it into very troubled waters.

Cavalli-Sforza had long been interested in human evolutionary history, and particularly the ways in which the study of genetic continuities and differences between geographically divergent and isolated populations could reveal the patterns of gene flow and human migration over the past two hundred thousand years. In 1991, convinced that this could be best explored through the study of the DNA of isolated groups, he proposed a study which, unlike the idealised universality of the HGP, would search for differences between these isolated populations.[15] He saw some of them as disappearing, making it all the more important that the work was carried out urgently.

The HGDP proposal soon came under criticism from both within and without science. Despite Cavalli-Sforza's unequivocal rejection of the concept of race, preferring the term population, which does not carry with it the concept of sharp boundaries and categorical genetic differences, anthropologists such as Jonathan Marks and Alan Swedlund argued that the proposal itself worked to remobilise the racial categories of the colonial past.[16] As no anthropologists had been included on the HGDP planning team, it was all too easy to point to the weakness of its core assumption that identifying an isolated population was a straightforward matter. Further, the critics challenged Cavalli-Sforza and his fellow geneticists' belief that the aboriginal peoples they proposed to study would, without demur, simply let the scientists take samples.

The geneticists belatedly saw the force of this argument and sought to recruit anthropologists to secure access; but what they singularly failed to recognise was that the anthropologists were already using DNA analysis in their own work on relatedness. It was thus unsurprising that the latter resented being cast as handmaidens in a study in which they would have expected to play a major part. There was also the scientific problem that the indigenous populations Cavalli-Sforza intended to study constituted only a tiny percentage of the world's population, and that if gene flow was to be studied it would be better done by studying more representative ethnically diverse populations in major urban centres – a position confirmed by the US National Research Council in 1997 in its review of the human diversity studies.

But the most dramatic challenge came from outside science, from the indigenous peoples (aided by NGOs) who, without being consulted, were to be the objects of study and whose blood or cheek swabs were to provide the research material. As Benito Machacca Apaza, President of the Community of Hatun Q'eros, a remote Peruvian tribe, put it: 'The Q'ero Nation knows its history, its past, present and future is our Inca culture and we don't need any so-called genetic study to know who we are. We are Incas, we always have been and always will be.'[17] In their focus on remote indigenous peoples, the HGDP planners failed to recognise their own identity as EuroAmericans, and that their planning was taking place in a particular country with a particular history of genocide and slavery. The now politically mobilised descendants of those who had so grievously suffered, however, did. The Native Americans were now part of an international movement of indigenous people in mutual solidarity with the isolated others of the HGDP.[18] Native Americans were engaged in political and economic struggle with the descendents of the colonists who had stolen their land. Their successful land claims rested on the narrative of their ancestors as the original inhabitants, who collectively not individually owned the land. Their narrative of origins was now under direct threat from the geneticists' rival origin narrative – one of migration across the Bering Strait and down through Alaska. What for the evolutionary geneticists would constitute a scientific success, would for the tribes threaten to delegitimise their hard-fought land rights.

The focus on indigenous peoples meant that the HGDP was initially uninterested in the genomes of the descendants of slavery. That was left to the geneticists at Howard, the pioneering African American university, using gene markers and SNPs to identify the locations – even particular villages – in Africa from whence the slaves had been taken. However, as the indigenous people's opposition intensified, self-identified African Americans attacked the HGDP for its racist exclusion of their DNA. To fend off the criticism, the project reversed its exclusion, and began gathering the merged DNA history of white slave owners and African women slaves. In so doing, however, Cavalli-Sforza's commitment to the concept of population genetics as requiring the study of inbreeding isolated populations, as opposed to racialised genetic differences, was fatally undermined.

At its inception the HGDP was interested in DNA derived from members of small groups, such as those living in the depths of the Amazon rainforest or remote areas of Papua New Guinea, whom they saw as isolated. But many of these allegedly isolated groups found ways of joining the opposition to the HGDP – an opposition based on their experience of the prior wave of green biopiracy in which Western biotechnologists had stolen and patented the knowledge of indigenous healers concerning the therapeutic properties of local plants and trees. The opposition saw the HGDP as amounting effectively to a second wave of biopiracy, this time aimed at the blood of the peoples themselves. As Jonathan Marks observes, to an anthropologist blood is never just blood. Mere research material to science, blood can at the same time have symbolic value, especially in non-scientific cultures. Above all there was anger, even fury, among the indigenous groups that the researchers appeared to care more about their DNA than their fate as disappearing peoples. The conservation of threatened species, from pandas to butterflies, seemed to rank higher than that of their own. As exchanges between the Native American tribes and the staff of the HGDP developed, there was a greater understanding on both sides, but the gap between the project and the opposition proved to be unbridgeable.

Faced with intense ethical and political hostility, public funding for the HGDP was discontinued. However, Cavalli-Sforza was able to secure agreement with CEPH in France to collect and store existing cell lines drawn from distinct population groups, thus avoiding the hostility to new collections. The project, now called simply the Genogeographic Project, was reconstituted with private funding from, among other sources, *National Geographic* magazine, which advertises on the internet for volunteers to provide samples that can be sequenced to reveal a person's ancestry. Manifestly, such donors were not likely to be drawn from the isolated groups of Cavalli-Sforza's imaginary.

It was not until 2001 that a new international publicly funded diversity project was proposed, the International HapMap Project (IHMP), but this time under much stricter ethical control. Once again the plan was to study diversity by way of variations in the non-coding regions of the genome, particularly SNPs. Many SNPs are located close together in the genome and are passed down the generations as a group within each chromosome. These groups are called haplotypes. Thanks to this linkage it is not necessary to genotype all the estimated 10 million SNPs per human genome in order to look at variations between populations – a mere 500,000 within haplotypes will suffice. Rather than following the HGDP's approach of looking at isolated populations, the IHMP collected samples from self-identified Han Chinese, Yorubas from Ibadan, Nigeria, Japanese from Tokyo and a group of European origin in Utah (the Mormon interest in genealogy helps explain this latter choice). Despite the much stricter ethical guidelines proposed for the project, Native Americans again refused to collaborate, as HapMaps still had the potential to threaten their land rights.[19]

AFTER THE HGP

Sequencing the 3 billion nucleotides of the human genome was a triumph of technological ingenuity, dependent on, and in its turn fostering, advances in automation and computing (Venter says that at one point during the race he had access to the greatest assemblage of non-military computer power in the world). The HGP came in ahead of schedule and under budget, and yet such is the rate of innovation that its timescale and costs now seem to belong to the remote past. The acceleration has more than matched Moore's well-known law of the two-year doubling time of computer power. Successive generations of massively parallel sequencers and dramatically enhanced data processing meant that within a decade of the Clinton–Blair press conference, anyone with a fat wallet could have their entire genome sequenced in a matter of weeks for around $10,000. The speed-up continues: by 2012 the $1,000 genome was in sight, and some companies are predicting that the sequencing will soon be capable of being done within hours for a hundredth of that price. New projects are underway; one to fully sequence a thousand genomes; a second, the human variome project (HVP), intended to uncover all the genetic variations associated with human diseases, was launched in 2010, based in Australia and largely bankrolled by China.

The HGP has been a wealth creator, even though the biotech companies have been hard put to turn a profit from the project. According to a study carried out by the Battelle Institute, over the fifteen years to 2010, the US government spent $3.8 billion on the HGP and related projects, which in turn added $796 billion to the US economy and supported 310,000 jobs.[20] This massive sum includes $244 billion in personal income. The analysis, commissioned by the biotech company Life Technologies Foundation, does not claim that this economic impact came entirely or directly from genomics, but only that the new jobs were born of and are still driven by the technologies developed by the HGP, as well as those developed by the private effort led by Celera.

One aspect of this wealth creation has been the move into 'retail genomics', perhaps the equivalent in the human life sciences of Henry Ford's Model T. In 2007, a California-based company, 23andMe, began offering personal genome scans based on a saliva sample – 'spitomics' to the more irreverent. The scan doesn't provide a full sequence, but, according to the company's prospectus, will identify the presence or absence of selected risk-bearing genes for up to 100 diseases – late onset conditions like heart disease, diabetes, and some cancers. It will also predict more cosmetic features and trace ancestry – all foras little as $199 (the price in 2011). 23andMe's venture was soon to be followed by several rivals, including deCODEme, a spin-off from the US–Icelandic company deCode, whose fate we trace in Chapter 5. The 23andMe product is sold on the basis that it will 'empower the purchaser and accelerate research'. In an uncertain financial climate the marketing of personal genomes has become a means by which several such biotech companies are attempting to secure their futures. However, the reliability, reproducibility and utility of the tests has been widely criticised. Martin Richards, a dark-haired Cambridge psychologist with extensive experience of working on the social aspects of human genetics, described to an audience of amused colleagues how his test report had informed him he had fair hair. More seriously, a study by a group from the Erasmus University in the Netherlands reported that the tests gave wildly inaccurate predictions, which also varied depending on which company performed the analysis. A similar discordance for some of the tests – notably on his risk for prostate cancer – was described by Francis Collins, who sent his own samples to three different companies and compared the results. But the prostate test is notoriously unreliable, as there is no one gene associated with the risk (at least 157 gene variants have been described). Even where they did agree, the advice on risk factors that Collins received – that he should diet and exercise sensibly, eat more oily fish, etc. – would in any case already be familiar to any well-informed upper-middle-class American, whether a biomedical researcher or not. The prostate example illustrates another problem with such tests – there is no evidence that early intervention is of any help in preventing the disease.[21] Such criticisms have led to calls to regulate or even ban the marketing of these tests, and have not helped the 'spitomics' companies navigate the stormy financial waters of the last few years. 23andMe ran into serious difficulties and has currently found shelter under the spreading mantle of Google. Nonetheless, the market in personalised genomics seems set to expand. How long will it be before the 'genome on a CD' envisioned by Gilbert becomes a normal part of the family photo album, just as nowadays the foetal ultrasound scan is the first picture in the baby book?

Despite the scale of these developments, there is a widespread recognition that the hyperbolic claims for the health and welfare benefits of the HGP are far from being proven. More genetic risk factors identified, more probabilistically predictive tests, yes, but gene therapy or personalised medicine based on genetic information, no. As for Koshland's wilder hopes, no comment is required. Why have these vast public subsidies not yet yielded the hoped-for results in health, or even medicine, as opposed to the optimistic wealth calculations of the Battelle Institute? The tenth anniversary of the publication of the genome was the occasion for some sober reflections. According to the US National Human Genome Research Institute, the impact of genomics on health care will only begin to build after 2020. James Evans, editor of *Genetics in Medicine*, states 'we need to quit trying to push genetics into medicine … we hear these grandiose statements that genomic technology is going to revolutionise medicine, but if so, it is going to take decades'.[22] It's time, he says, to deflate the genomic bubble.

The failure, and some of the reasons for it, are well summed up in a paper by molecular geneticist Eric Lander, one of the key figures in the history of the HGP, in a celebratory retrospective published in *Nature*:

> Our knowledge of the contents of the human genome in 2000 was surprisingly limited. The estimated count of protein-coding genes fluctuated wildly. Protein coding information was thought to far outweigh regulatory information, with the latter consisting largely of a few promoters and enhancers per gene. The role of non-coding RNA was largely confined to a few classical cellular processes ... A decade later we know that all of these statements are false. The genome is far more complex than imagined, but ultimately more comprehensible because the new insights help us imagine how the genome could evolve and function.[23]

Nonetheless Lander remains optimistic that despite the – at best – slow to arrive health benefits of the past decade, geneticists have learned the right lessons, that a better understanding of common diseases, from Type 2 diabetes to cancer, kidney disease and psychiatric disorders, will be gained, and that genomic medicine is indeed moving 'from base pairs to bedside'.

This belated recognition of what was not acknowledged a decade earlier would, however, come as no surprise to those biologists who had long argued that molecular geneticists were ignoring the rest of biology – notably evolution, Darwinian natural selection and development – at their peril. Leading evolutionary and population geneticists such as Ernst Mayr, Francisco Ayala, Stephen Jay Gould and Richard Lewontin, developmental biologists like Anne McLaren, and even sociobiologist E.O. Wilson, have long been critical of this kind of molecular reductionism which conceives the overall project of the life sciences as one of disaggregation – breaking down nature into ever smaller parts, explaining higher level phenomena, such as development or behaviour, in terms of lower level sciences such as biochemistry. It is not so much that genomics has, as Michel Foucault would have it, extended and intensified the biomedical gaze, but rather, as we will explore in the next chapter, that the project of disaggregation has lost sight of the organism itself, let alone of its connectedness with the entire socio-ecological system over time(in terms of both the lifecycle of a given individual and the evolution of the species as whole). Until the findings of the HGP can be integrated within these perspectives, whatever wealth it may bring in its train, the wider health benefits will remain at best elusive, and at worst suffer from the diversion of funds that such molecularisation entails.

Nevertheless, the standard molecular biology textbooks continue to neglect these 'higher levels' of biological organisation. Lander's reference to evolution contrasts with Watson's (et al.) immensely influential textbook *Molecular Biology of the Gene*, which mentions Darwin just once – in a chapter entitled 'The Mendelian View of the World' – and natural selection not at all. Molecular biologist Bruce Alberts's (et al.) text *Molecular Biology of the Cell* mentions natural selection again just once, in a 1,600 page book. The tendency to ignore evolution in the way that both these textbooks do has shaped the reductionism – both as method and philosophy – of several generations of molecular biologists. From Crick and Watson onwards, the supremely confident belief of the molecular biologists was that disaggregating the organism, disaggregating the cell, would deliver knowledge of bios, of life itself – a move only to be compared with that from atomic to sub-atomic physics. Their programme, which put DNA at its core, won the day, not only among the scientists, but through them the politicians and venture capitalists, the research foundations and Big Pharma, who delivered the resources crucial to the project of the molecularisation of life.

It is not merely that the HGP relied for its success on molecular and information technology. Depending as it does on an intensely reductive view of living organisms, a molecular and informatic view of the nature of life is embedded in its very conception. Humans, like other creatures, are to be understood by reducing them to their molecular constituents. These constituents, above all DNA, are described as informational macromolecules. Occurring contemporaneously with the growth of molecular biology, the development of computer technology, with its demands on information theory, has not merely provided the technical instrumentation and computing power. It has also provided the organising metaphors within which the data is analysed and the theories created. As is so often the case, it is Richard Dawkins who puts it most clearly: life, he insists, is digital, not analogue. Crick may have coined the informational metaphor, but it is Dawkins who has taken it to its logical conclusion. Consider, for example, the euphoric elegance of the passage in *The Blind Watchmaker* in which he considers a willow tree in seed outside his window: 'It is raining DNA … It is raining instructions out there; it's raining tree-growing, fluff-spreading algorithms. That is not a metaphor, it is the plain truth. It couldn't be any plainer if it were raining floppy discs.'[24]

Had he been writing a decade later, in the 1990s, it would have been raining Gilbert's CDs. This is stylish stuff, great fun to read, so much so that it has found its way into anthologies of scientific prose. Yet for the reader who prefers to unpick rhetoric the sentences are problematic. To declare that a metaphor is not a metaphor and is nothing less than the plain truth smacks of the kind of quasi-religious certainty that Dawkins elsewhere finds so objectionable. Crick's Central Dogma was pronounced with a certain judicious irony, but Dawkins doesn't do irony. Instead he insists on the plain truth of The DNA Delusion.

The price of this delusion, for biological theory and for the hopes and hypes of the HGP, will become clear in the next chapter.

2

Evolutionary Theory in the Post-Genomic Age

MATERIALISM AND *THE ORIGIN OF SPECIES*

As the geneticist and sequencer Eric Lander realised, the limits to what the HGP might reveal about human nature, either in general or specifically, have been strikingly exposed over the decade since the sequence was announced. Francis Collins found little of value from the list of risk factors he received when he sent his DNA samples to 23andMe. The paradigm that had framed genetics since the 1950s was beginning to fall apart. Unless set in the context of the development of the organism, DNA sequences were meaningless, and unless set in the context of evolution, the fact that humans and chimpanzees have nearly identical sequences was incomprehensible.

However, it was not a scientific revolution that was required, but a recovery of those central preoccupations of the life sciences which had been lost in twentieth-century molecular enthusiasms. To this extent at least, although Cavalli-Sforza's project was both resisted by his proposed subjects and sharply criticised by anthropologists, his focus on correlating variations in human DNA sequences with migratory history and changing cultures had recognised, and attempted to respond to, precisely this loss. In the nineteenth century biologists understood the heredity and development of the organism as two facets of a single science. Darwin's achievement was to embrace both within his theory of evolution. It was the rediscovery of Mendel and the birth of modern genetics early in the twentieth century that separated heredity from development, each with distinct conceptual frameworks and different methodologies. The complexities of the genome revealed by the HGP and its puzzles provided the impetus for reintegrating these separated siblings, just in time for the new millennium, under the catchy new name 'evo-devo'.[1]

To understand the relevance of evo-devo we need to go back briefly to the time before their separation, to the success of *The Origin of Species*, published in 1859 in the cultural context of a growing physicalist materialism running alongside the religious certainties of Victorian Britain – without, for the most part, causing too much trouble. It isn't possible to understand and interpret the construction of human nature by genomics without locating it in the historical tradition of this materialism, voiced with increasing cultural authority since the birth of modern science in the seventeenth century. *Nullius in Verba* was the challenge issued to philosophers by the Royal Society at its foundation, claiming that only the new knowledge of science – with its commitment to experimentation and its systematically gathered evidence presented preferably in mathematical form – had the capacity to produce truth. In doing so the new sciences made the powerful move of offering an alternative materialist account to the creation stories around which Western culture had been built.

By the late eighteenth century the German Franz Gall's phrenology was attempting to locate mental attributes, and the Italian physiologist Luigi Galvani's 'animal electricity' life itself, within the explanatory realm of the natural sciences. In distant Cornwall, the young chemist Humphry Davy and the poet Samuel Coleridge formed a lifelong friendship, and Davy's speculations about electricity as a life-force lay behind Mary Shelley's unforgettable creation of Frankenstein's monster. Her book, first published in 1818, was however profoundly critical of the amoral character of science and scientists; she saw the horrors committed by the monster as arising not from his physical appearance but because his creator denied him love – a criticism better recognised by feminists than by mainstream science-fiction analysts. These materialist accounts of nature and human nature produced by natural philosophers (renamed scientists in the 1840s by the Oxford philosopher William Whewell, an inveterate spinner of neologisms) found a receptive audience among

Victorian intellectuals, then a relatively small group sharing a common culture. High up in the Yorkshire Dales, the Brontë sisters (their brother Bramwell was probably in the pub) would walk the several miles from the Haworth parsonage down to Keighley, their nearest town, to listen to a lecture on phrenology as the hottest materialist account of brain and mind. In Charlotte Brontë's vivid depiction of the head shapes of Mr Rochester and Jane we read the new phrenology in the pen of the writer.

Shelley's *Frankenstein* became a best-seller, but could not hold back the fully reductive materialism that was taking a firm hold within the life sciences. In 1845, four rising German and French physiologists, von Helmholtz, Ludwig, du Bois-Reymond and Brucke, swore a mutual oath to prove that all bodily processes could be accounted for in physical and chemical terms. The Dutch physiologist Jacob Molcschott defended the position most strongly, claiming that 'the brain secretes thought like the kidney secretes urine', while 'genius is a matter of phosphorus'.[2] For the zoologist Thomas Huxley, mind was an epiphenomenon, like 'the whistle to the steam train'. But it was above all Charles Darwin who provided the intellectual and empirical bedrock for a materialist account of human origins and human nature. Although evolutionary ideas were far from uncommon, it was *The Origin* that, by providing an intellectually satisfying and coherent secular origin-story of life itself, was recognised as precipitating and symbolising a transformation in Western culture. While an outspoken minority, like his cousin Galton or disciple Huxley, embraced atheism, many, like Darwin himself, though a materialist in his outlook, continued to accompany his wife Emma and their children to church (although after the death of their beloved daughter Annie, he went no further than the church door). Darwin's radical secularity was confined to his intellectual life; socially he was entirely conventional. Odd though it may seem to today's sensibilities, his burial in Westminster Abbey was of a piece with the contradictory currents of nineteenth-century England.

Certainly biologists had been questioning the notion of the fixity of species for a good three quarters of a century before *The Origin*, but they were unable to find a convincing theory of evolutionary change. In his autobiography Darwin writes of his almost accidental reading of Malthus and of his epiphany as he understood that 'I had at last got a theory by which to work'.[3] This was Malthus's doctrine of the inexorable natural growth of human populations until they outstripped the supply of food, with its political correlate that the weakest must be permitted to go to the wall, 'applied' as Darwin wrote, 'to the whole animal and vegetable kingdoms'. Where Malthus politically endorsed the weakest going to the wall, Darwin depoliticises it, making going to the wall a law of nature. The Darwinian thesis is thus straightforward: (1) in an environment of limited resources, all organisms produce more offspring than can survive to adulthood; (2) although offspring resemble their parents, there are minor variations among them; (3) those variations better adapted (fit) to their environment are the more likely to survive and reproduce in their turn; (4) thus such favourable variations, if inherited, are likely to be preserved in subsequent generations. It was the prominent English polymath Herbert Spencer who described this process as 'survival of the fittest', directing attention to the relevance of Darwinian theory to nineteenth-century capitalist and imperial Britain. However, Darwin's fellow biologists remained uneasy with the theory because, in the absence of the concept of genes, there was no obvious mechanism for how such better adapted variations were transmitted to subsequent generations. Darwin, well aware of this problem, with which he wrestled for the rest of his life, was reduced to speculating that the hidden determinants responsible for the transmission of fitter forms were carried in 'gemmules' in the bloodstream.

Apart from the reference to Malthus, in *The Origin* Darwin fails to acknowledge his indebtedness to the work of others. There is no mention in the first five editions even of his grandfather Erasmus's romantic view of evolution, echoing the Swedish naturalist Linnaeus, the great classifier of species, for whom even plants share love. Nor of his eminent French predecessor, the zoologist Jean-Baptiste Lamarck's understanding of evolutionary change occurring through individual effort – or the several evolutionary currents that had flowed through the debates of the first half of the nineteenth century; instead the steady emphasis is on 'my' theory. Yet Darwin's library, his friendships and immense correspondence reveal that his work was profoundly influenced by many natural philosophers. Not until the final edition of *The Origin* in 1872 did he redress the criticism made by some reviewers that he had erased his predecessors, by adding as a preface an 'historical sketch'.

The introduction to the first edition does however make a courteous acknowledgement of Alfred Russel Wallace, who had 'arrived at almost exactly the same general conclusions that I have on the origin of species'. Wallace, working as a specimen collector in the Malay archipelago, had sent his manuscript to Darwin for publication, precipitating the latter's panic over being pre-empted. The botanist Joseph Hooker and the geologist Charles Lyell, Darwin's long-standing and highly influential friends, ensured the simultaneous reading of Wallace's paper and a matching one from Darwin at the Linnaean society in 1858. Wallace's claim to priority was gently eased aside, and the socially weaker man, rather than contesting the more powerful Darwin, expressed only his gratitude and deference. Wallace's paper pushed Darwin into a frenzy of writing, completing the 'abstract' (400 pages) that comprised *The Origin* just a few months later. Wallace's socialism and proto-feminism were politely but firmly erased.[4]

Marx and Engels in their commentary on Darwin spoke of his materialism as 'mechanical', to be distinguished from their own dialectical and historical materialism. The supernatural, and hence religious belief, was simply redundant in biology's account of human nature. Within the very different materialism of Marx, religion is theorised as 'the sigh of the oppressed creature, the heart of a heartless world, just as it is the spirit of a spiritless situation. It is the opium of the people.' It is difficult to think of two materialisms further apart. Writing to Engels some three years after the publication of *The Origin*, Marx, recognising Darwin's dependence on Malthus, anticipates a central premise of science studies – the mutual shaping of science and society:

> It is remarkable how Darwin rediscovers, among the beasts and plants, the society of England with its division of labour, competition, opening up of new markets, 'inventions' and Malthusian 'struggle for existence'. It is Hobbes' *bellum omnium contra omnes* and is reminiscent of Hegel's Phenomenology, in which civil society figures as an 'intellectual animal kingdom', whereas, in Darwin, the animal kingdom figures as civil society.[5]

This is not how mainstream biologists read Darwin's theory of evolution. Instead of problematising his reproduction of the Victorian social order within *The Origin*, they read out his embrace of capitalist political economy and the sexism and racism which for him were integral to nature's laws. For mainstream biologists the true reading is neutral, having nothing to do with the social order. It is the meticulous study of the natural order and the illumination that evolutionary theory casts upon it. Since for biologists humans are part of this natural order, the theory that applies to non-humans applies to them as well. In the intervening century and a half, biologists have continued building on Darwin by insisting on a materialist account of nature in general and of human nature in particular, from our basic physiology to our powers of cognition, our emotions and beliefs. As we will discuss in later chapters, the current project of the neurosciences to explain not only the brain but the mind, and indeed consciousness itself, seeks to realise Darwin's fundamental premise.

DARWIN'S LEGACY: THE TREE OF LIFE AND THE RACIAL AND SEXUAL HIERARCHY

The first edition of twelve hundred copies of *The Origin* sold out within days of publication, as the second edition was already being prepared, though these sales were small beer by comparison with Herbert Spencer's, that ultimate ideologue of nineteenth-century capitalism, whose books sold more than a million copies, mainly in the US, within his lifetime. No longer were 'kinds' to be seen as created independently in the brief interlude between God's separating heaven and earth and then resting on the seventh day.

Rather than having been intelligently designed, as the Reverend William Paley had famously insisted half a century previously, for Darwin species had evolved from a single common origin and been transformed by selection operating on random variation over long periods of geological time. Darwinian evolution was non-teleological – there was no divine purpose, no foreseeable future. Darwinian evolution, according to the astronomer John Herschel, was merely 'the law of higgledy-piggledy'.

The book's resonances were widely shared and despite religious objections evolutionary theory was becoming part of the wider culture. For Spencer, Darwinian natural selection provided the explanation for why laissez-faire liberalism required a brutal 'struggle for existence'. Darwin, despite regarding Spencer's work as speculative, later adopted the term – and even later regretted having done so. Had he instead adopted the Russian anarchist and biologist Prince Peter Kropotkin's 'struggle for life' in which mutual aid, not mere existence, was a driving force in evolution, the naturalising gloom of Darwinism might have been averted. The proposition that it is not only competition but cooperation – both within and between species – which has been a major driving force in evolution, has run like a subterranean heresy through most of the evolutionary theorising of the last century. Only now – as we discuss later in this chapter – has it re-emerged into the daylight of mainstream thinking.

For Darwin, evolution is an ongoing process without endpoint. Although natural selection rejected the theological view of the Great Chain of Being – which informed the classificatory system proposed by Linnaeus, in which all living organisms were ranged in a God-ordained hierarchy – evolution was still seen as progressive, with lower organisms giving way to higher ones. Darwin represented this as a many-branched tree of life, with *Homo sapiens* at the highest point. (Today's evolutionary biologists prefer the metaphor of the bush, with all currently extant species equally 'evolved'.) And yet despite his insistence that natural selection has no goal or end point he still remains something of a nineteenth-century social progressivist, and speculates in the closing pages of *The Origin* on the wonderful civilisation of the future as the species evolves. 'And as natural selection works solely by and for the good of each being, all corporeal and mental endowments will tend to progress towards perfection.' Later evolutionary theorists from the philosopher Henri Bergson to the Catholic palaeontologist Teilhard de Chardin were, however, to reassert the evolutionary teleology that the Anglophone tradition rejected.

The Origin only hints at the relevance of the theory to humans. However, within a decade of the first edition, in 1869, Darwin's cousin, the pioneering biometrician Francis Galton, had published *Hereditary Genius*, a foretaste of his eugenic proposals which were to haunt the twentieth century (see Chapter 4). Darwin welcomed Galton's ideas, drawing on them two years later for his most provocative book – *The Descent of Man and Selection in Relation to Sex*. It was not until *The Descent* that Darwin finally affirmed humanity's ape-like origins, locating human differences within an evolutionary framework. To enrich his argument he spent much time studying the behaviour of primates, notably the solitary orang-utang in the London Zoo. In the twentieth century, primate behaviour in the wild began to be studied systematically, perhaps best known through Jane Goodall's many years of ethological observations of the chimpanzees of Gombe in Tanzania, documenting the variety, richness and complexity of primate societies in their natural habitats. Despite the development of ethology with all its promise, the practice of inferring 'normal' ape behaviour by observing them in captivity persisted. The eminent anatomist Solly Zuckerman, both chief government science adviser and President of the London Zoo, was one such, drawing conclusions about the innate violence, sexual practices and hierarchical structures of primate societies through observing their social interactions in cages.

Unlike a number of his contemporaries, Darwin embraced a monogenic view of the origin of the human species. Certainly *The Descent* divided humanity into many distinct races, describing their differences in skin, eye and hair colour in some detail. But Darwin nonetheless insisted that there was a single human origin, the various human races/variants having separated from this common stock over evolutionary time. This difference from the prevailing polygenic view in which each race had a distinct origin was a major issue for biological theory. The concept of race was entirely unproblematic for all the participants in these debates; it wasn't until a century later that most prominent human population geneticists insisted that 'race' had no meaning or usefulness for human biology. Challenged by the renewal of scientific racism in the 1970s, many geneticists insisted that this was mere pseudo-biology, seeking to claim the authority of science while ignoring its procedures. Scientifically unsustainable, scientific racism was socially toxic.

Like most Victorian gentlemen at the height of Britain's imperial power, Darwin shared the belief in a racial hierarchy ranging from the less evolved, degraded savages of Tierra del Fuego – whom he had seen on his long voyage on the *Beagle* in the 1830s – to the higher European civilisation, not least that of his home, Down House in Kent, the garden of England. Indeed, he went further, arguing that the evolutionarily inferior black races would in due course be out-evolved and defeated by the whites. Despite his monogenic perspective, he remained trapped within a very nineteenth-century view of fixed racial and sexual hierarchies. Thus, while his hatred of slavery was intense, his concept of race essentialises difference, so that variation within the species slides into a hierarchy between the races.

Sexual selection is almost as central to Darwinian evolution as is natural selection, because it both explains the differences between the sexes within a single species and some of life's extreme and otherwise apparently non-adaptive features such as the glories of the peacock's tail. Darwin argued that sexual selection accounts for the fact that males and females of the same species often differ in shape and size. Males compete for females; they may fight like stags, or display like peacocks. Females then choose the strongest or most beautiful male. This, he claimed, ensures the reproduction and selection of those male characteristics that females find most attractive; thus within the species it is only males that evolve in order to meet nature's criteria of strength and beauty.

Darwin struggled with the idea of beauty – easy to see in the peacock's tail but less so in the turkey-cock's wattle. The problem is encapsulated in the commonplace that beauty is in the eye of the beholder – turkey hens may have a different view from the observing human. Modern-day biologists substitute ideas of beauty with the claim that flamboyant male appendages signal 'good genes'. Although sexual selection is regarded as one of the core features of evolutionary theory, and popular writing, especially from evolutionary psychologists, accepts it unquestioningly, attempts to demonstrate it empirically have not proved entirely successful. Furthermore, there is evidence that both sexes have other potential sexual strategies. For example, while massively antlered stags fight, females may choose to mate off-stage with less well antler-endowed males.

When he turned to humans, Darwin's view of the differences between men and women was entirely of his time. For his part, Wallace never accepted sexual selection, and he was far from easy with Darwin's extrapolation from apes to humans; he wanted to leave space for the soul. Darwin however was unequivocal; as he writes in *The Descent*, the result of sexual selection is for men to be 'more courageous, pugnacious and energetic than woman … [with] … a more inventive genius. His brain is absolutely larger … the formation of her skull is said to be intermediate between the child and the man'. As a consequence, 'man has ultimately become superior to woman', and it is a good thing that men pass on their characteristics to their daughters as well as to their sons, 'otherwise it is probable that man would have become as superior to woman, as the peacock is in ornamental plumage to the peahen'.[6] Here Darwin seems to be influenced by Galton, who viewed genius as entirely passed through the male line, with women as mere empty vessels. Darwin shared his cousin's androcentricity – and this was not missed by contemporary feminist intellectuals. Within five years of the appearance of *The Descent*, the American feminist Antoinette Brown Blackwell,[7] even while welcoming evolutionary theory, complained at Darwin's erasure of women. Well aware that as a woman she had no training as a biologist she looked to the future when feminist biologists would return better armed to the fray. As we will see, it took a hundred years.

THE MODERN (OR NEO-DARWINIAN) SYNTHESIS

By the early twentieth century the concept of evolution had entered the intellectuals' common culture, but among biologists the theory of natural selection was in considerable difficulty. For all his decades of research, Darwin had been unable to provide a mechanism whereby favoured changes could be preserved across generations. When Gregor Mendel's studies were rediscovered in 1900 after decades of neglect, they suggested a mechanism for evolutionary change that had eluded Darwin. The almost simultaneous rediscovery and recognition of the significance of Mendel's studies marks the birth of genetics – the term coined for the new science by William Bateson. Mendel referred to the internal factors he postulated as transmitting the colour and shape of his peas as 'hidden determinants'. It was the Danish plant biologist Wilhelm Johannsen who called them genes. Genes, it was discovered, could mutate. This offered an explanation for the emergence of new varieties – even new species – and hence for evolutionary change. In the absence of mutation, genes were immortal, immune to somatic changes, the unmoved movers that determined all bodily functions. It was not until the 1930s that the Mendelian and Darwinian mechanisms were reconciled, in the work of mathematically minded geneticists J.B.S. Haldane and Ronald Fisher in England and Sewall Wright in the US. They brought together genetics and natural selection as the accepted mechanism for evolution, called the Modern Synthesis by Haldane and Fisher and neo-Darwinism by Wright. In this synthesis, it was genes that carried the variations within a species from generation to generation. By the 1950s the formal definition of evolution had become 'a change in gene frequency within a population'. Organisms had disappeared from the account; what mattered was not even the genome but individual genes operating independently of one another – an approach derided by Sewall Wright as 'beanbag genetics'. Unlike the mathematicians, his real-world field studies made him more aware of complexity, and of the interaction of genes during development.

However, beanbags won out for mainstream genetics, and this

privileging of genes over and above the organisms in which they are embedded has had a disastrous consequence for the life sciences. The genetic turn deprived the term 'evolution' of one of its original, pre-Darwinian meanings – development, the unrolling lifecycle of any living creature. Developmental biology's starting point is the study of similarities – the biological processes that generate all living creatures, from the transitions between caterpillar and butterfly to the extraordinary uniformity in the way in which humans develop from fertilised egg through embryo, foetus, infant and into adulthood. Development thus emphasises form, pattern, wholeness and above all time, the species-typical dynamic of any organism's lifecycle. As ethologist Patrick Bateson and molecular biologist Peter Gluckman describe it, developmental processes must be simultaneously robust – capable of withstanding environmental insults – and plastic – capable of change in response to challenge.[8] Ignoring development enabled an inexorably reductionist genetics to become the study of differences between organisms, assumed to be coded in the genes.

Developmental biology, less amenable to the increasingly molecular interest of the geneticists, became by the 1930s the focus of concern for a group of self-consciously anti-reductionist (and after the rediscovery of Engels's *Dialectics of Nature*, 'dialectical') or 'systems' biologists, mainly based in Cambridge, among them the embryologist and later historian of Chinese science Joseph Needham,[9] mathematician Lancelot Hogben, mathematical biologist J.H. Woodger and developmental biologist C.H. Waddington.[10] However, lacking molecular tools their research programme was theoretical rather than empirically researchable, and was dealt a devastating blow when the Rockefeller Foundation, committed to a reductionist approach, rejected their proposal for a research institute in Cambridge based around their systems programme in favour of a major investment in what was to become molecular biology. Not until recent decades have the theoretical issues that occupied these 1930s biologists once again become an important focus of research, although stripped of the broader philosophical framing they had insisted on.

Meanwhile the synthesis of genetics and evolutionary theory was proving increasingly powerful, summed up by the population geneticist Theodosius Dobzhansky in 1950 as 'nothing in biology makes sense except in the light of evolution'.[11] From then on, and with increasing certainty following the discovery of the role of DNA as the genetic material, the triumph of neo-Darwinism seemed assured. Yet there are still major controversies over the processes of evolution and the mechanisms of speciation. Even the most basic issues – what it is that evolves, what adaptation is, and whether selection is the only motor of evolutionary change – remain in question. As we discussed in the last chapter however, these problems are almost entirely ignored both by molecular biologists and by those who wish to transfer the reductionist genetic ideas of the 1930s Modern Synthesis holus-bolus into twenty-first-century social sciences and humanities.

FROM THE MODERN SYNTHESIS TO THE NEW SYNTHESIS

What evolves? For Darwin and his immediate successors the answer was self-evident. It is organisms – or phenotypes, as they came to be called. However with the Modern Synthesis, attention turned from the organism to its genes. Evolutionary biology set aside field studies – the natural history of Darwin and Wallace – to focus on mathematical modelling. This emptying out of the organism reached its apogee in Richard Dawkins's description of genes as the active 'replicators' embedded in and controlling organisms, which are nothing but the passive 'vehicles' whose only function is to ensure the gene's transmission through generations.[12] Attractive as this thesis was to many at the time *The Selfish Gene* was published – perhaps because of its very simplicity – its elegance has been confounded by the complexity revealed by the molecular genetics of the twenty-first century. Despite this, Dawkins's support for genetic determinism is still influential, as evidenced by the continued popularity of his books.

But if it is genes rather than organisms that are important, any mechanism that perpetuates genes into the next generation will suffice – hence Haldane's apocryphal pub joke that he would sacrifice his life for two brothers (each carrying half his genes) or eight cousins. Haldane's joke was later mathematicised by William Hamilton as 'kin selection'. This provided a theoretical basis for E.O. Wilson's *Sociobiology: The New Synthesis* in which he extends evolutionary theory to include patterns of behaviour, and argues that natural selection favours behaviours that enhance the reproductive success of genetically related kin in social species.[13] It was Wilson's extrapolation of this thesis from ants – his research interest – to humans, and his reassertion of the Darwinian view of the natural divisions between the sexes, that created the furore. When a previous Harvard professor, Darwin's correspondent Louis Agassiz, had advanced similar arguments, the natural superiority of white males was seen as non-contentious on and off campus. But in 1975, when *Sociobiology* was published, second-wave feminism was at its height and hostility to the reduction of women to their biology at its fiercest.

No fewer than thirty-five fellow academics – including Wilson's Harvard colleagues, biologist Ruth Hubbard, population geneticist Richard Lewontin and palaeontologist Stephen Jay Gould – were members of the collective which rapidly published a rebuttal: *Biology as a Social Weapon*.[14] This charged Wilson with a crass genetic determinism that naturalised the existing hierarchies of power and control over resources between classes and genders and fostered racism. These charges were sufficiently serious that both books were included on student reading lists to ensure 'balance'. (Thirty years later, when Harvard President Larry Summers made similar remarks about the natural inferiority of women, he was forced to resign. Ivy League biosocial understandings of gender had changed under the pressure of feminism both as scholarship and as social movement.)

Just four years after the publication of *Sociobiology*, Ruth Hubbard challenged the androcentricity of Darwin's theory head on. In a pioneering essay she posed the question '*Have Only Men Evolved?*' At last a feminist and trained biologist was able to realise Brown Blackwell's hope. It was not that historians of science had ignored

Darwin's taken-for-granted social assumptions or had failed to question whether there was a significant difference between Darwinism and Social Darwinism, but Hubbard now turned the spotlight onto the ways which Darwin naturalised females and femininity, males and masculinity. His binary distinction with its stereotypical gendering has, she concludes, continued to play an unquestioned and damaging role within the life sciences.

A classical example of this binary thinking was the edited collection *Man the Hunter*,[15] published in 1968, wherein men were portrayed as the central providers of food, and the activities of women left at best under-reported. In the 1970s, with the changed consciousness born with the women's movement, feminist ethnographic accounts – with Patricia Draper's study of the !Kung perhaps central – instead reported that it was women who, in their gathering activities, were the main food providers, with men supplying the meaty treats. Adrienne Zihlman summed up this re-centring of anthropology with her challenge of *Woman the Gatherer*.[16] The challenge, backed by a powerful body of research, established a new consensus; today nomadic societies are termed Hunter-Gatherers, acknowledging reciprocity, not dominance. Feminist anthropologists were not alone in their evidence-based revision of the hitherto androcentric account of human evolution. Feminist primatologists, recognising the importance of apes in the narrative of evolution and that previous field studies were overly focused on males, turned to field studies to gather fresh evidence. Jeanne Altmann brought more systematic data-gathering methods from studies of children, bringing the activities of the female primates into focus. She was not alone; other influential feminist primatologists, notably Nancy Tanner, Alison Jolly and Linda Marie Fedigan, saw research on apes as critical for the reconstruction of the narrative of evolution. They were joined by sociobiologist Sarah Hrdy with her studies of Langur monkeys. To a remarkable extent these women had between them advanced the feminist project of de-centring science, making non-human and human females part of the story. This was not an easy task as the feminist movement of the '70s was largely hostile to the possibility of a feminist biology, seeing the socio-economic power of men, aided and abetted by the sexist ideology of science, as the main problem.

The significance of the study of primates as a battleground for the question of human origins was similarly recognised by feminist historian of science, Donna Haraway, in her influential *Primate Visions*.[17] For her, the scientists' account of nature reflects and constructs society and culture. Her deconstruction of the dioramas of natural history museums in the 1980s – still celebrating *Man the Hunter*, with women's activities restricted to cooking and child care – echoed the criticisms of the feminist ethnographers. But her deconstruction of primatology provoked anger among the feminist primatologists; as natural scientists they were committed to the replacement of inadequate accounts of nature with more adequate ones. Or, put more bluntly, to replacing error with truth. They saw Haraway's deconstructionism as undermining their science and as dismissive of their years of patient research. Haraway's self-defence – that she regarded all the accounts, including her own, as stories – did little to assuage the anger. Stories might be what historians produced, but not natural scientists. While this tension between the deconstructionist account of science and natural scientists themselves came to a head with the Science Wars of the mid-1990s, it was prefigured with a difference in the earlier feminist debates. The review of *Primate Visions* by primatologist Alison Jolly and her postmodernist daughter Margaretta exemplifies the tension: the mother hates the book, the daughter loves it, and both are feminists.[18]

EVOLUTIONARY THEORY AND THE DISAPPEARING GENE

Neither *Sociobiology* nor *The Selfish Gene*, nor their critiques, required reference to the molecular constituents of the gene or the biochemical processes by which they might exert their cellular effect. Wilson, Dawkins and their sociobiological colleagues continued to treat 'genes' as formal accounting units, irrespective of their materiality as DNA. At the height of gene talk in the 1980s and '90s, genes were seen as having the capacity to explain everything – there could be beanbag genes 'for' altruism, sexual preference, bad teeth or anything you wanted – which could be fitted into a model of evolutionary change. As Gould and Lewontin caustically observed, these were nothing but just-so stories.

Approaches that treat the gene as a theoretical unit were and remain astonishingly indifferent to the new molecular genetics, clinging instead to the 1953 conceptualisation of genes as lengths of DNA providing the template from which proteins, and hence cells and organisms, could be synthesised. A mutation in a gene (a substitution or deletion of one or more of the nucleotide letters in the sequence) would, it was assumed, change the structure of the protein for which it coded, and hence by a long chain of effects would result in a change in phenotype on which selection could act. The research of the molecular labs seemed for a time to coincide perfectly with the predictions of the evolutionary theorists. It was this picture that sustained the genetic myth of DNA as akin to God, the unmoved mover, the information carrier and controller of cellular processes, the modern avatar of the immortal, eternal germplasm of the early 1900s.

But as Eric Lander ruefully pointed out, this simple molecular picture was too good to be true. Over the subsequent half-century the initial elegant clarity has been made vastly more complex. Human cells contain up to 100,000 different proteins, but the HGP showed that there are only something over 20,000 protein-coding genes – about the same number as in the fruit fly Drosophila. More than 98 per cent of the three billion As, Cs, Gs and Ts in the genome do not code for proteins at all. Some have important tasks in regulating the timing at which the coding genes are switched on, but, as we mentioned in the last chapter, much has been called disparagingly 'junk DNA'. But just how much of it is really junk and how much has currently unanticipated functions will depend on the results of the ongoing attempts by Venter and others to build synthetic organisms by inserting artificially constructed DNA sequences into living cells.

A further complexity arises because those strands of DNA that do code for proteins are not arrayed in a continuous sequence, but scattered in bits amid other non-coding sequences. Today, when molecular geneticists report the discovery of a 'gene for' longevity or obesity, they are using distinctly confusing language. They are referring, more precisely, not to 'a gene' but to a group of DNA sequences, stitched together and activated by cellular mechanisms during development, which to a greater or lesser degree enhance the probability of the individual carrying the sequences living longer or becoming obese. The gene of the molecular biologists is thus very different from the 'accounting unit' gene of the evolutionary model builder. For molecular geneticists therefore, the idea of 'the gene' as the fundamental unit of life, akin to the early twentieth-century physicists' conception of the atom, and just as powerful in its time, was long past its sell-by date, as Evelyn Keller argues in *The Century of the Gene*.[19] Despite this, the discourse of evolutionary theorists has remained that of genes, not DNA, as for them molecular mechanisms are irrelevant – even an obstacle to grand theorising. This twin track persists, and although unquestionably the bulk of the research monies goes to the molecular biologists, the book sales by the sociobiological modellers, and their impact on popular culture, are impressive.

GENES AND DEVELOPMENT: EPIGENETICS

The reconceptualisation of genetics that has followed the sequencing of the human genome has made it clear that the cellular regulatory processes controlling when and which genes are activated are crucial to development. Rather than DNA determining cellular activity, it is the cell in which the genome is embedded that 'chooses' which bits of DNA to use to build which proteins, and when and how, during the developmental sequence that leads from fertilisation to the adult organism. This process is known as epigenetics, a term coined by Waddington in the 1950s, and currently one of the hottest fields of molecular biology. Epigenetics addresses two major problems. First, how can 20,000 genes, each of which is present in every cell of the human body, be used in the ontogeny of some 250 different cell types, each with a characteristic structure and function, and each containing a different subset of the 100,000 proteins? To add to the complexity, the different cell types are 'born' at different times during the ordered development of the foetus and must migrate to appropriate regions of what will in due course become the fully formed baby. Second, how can seemingly minor environmental events at key stages in development produce a cascade of changes in the developing organism? The Thalidomide catastrophe of the late 1950s – when the drug was extensively prescribed in Europe to pregnant women to counter morning sickness, on the assumption that it could not cross the placental barrier – cruelly illustrated the price of ignoring such a minor environmental intervention. In the US, Frances Kelsey at the Food and Drug Administration judged the clinical trials inadequate, and refused the drug a licence, thus protecting American women and sparing their babies from the deformities the drug produced.

Epigenetic studies are uncovering a dazzling array of regulatory processes by which signalling molecules – sometimes themselves proteins, sometimes small molecules, some generated internally by each cell, some diffusing from other regions of the developing foetus – act as switches, turning particular stretches of DNA on or off so as to ensure that particular proteins are synthesised at the appropriate moment in the developmental sequence. Alterations in the timing of these switches may result in huge changes in the adult phenotype, producing new variations on which evolution can act. Genes are no longer thought of as acting independently but rather in constant interaction both with each other and with the multiple levels of the environment in which they are embedded. Interpreting what some are calling the 'epigenome' thus represents a return to the systems or dialectical research programme of Needham, Waddington and their followers, but now armed with the tools of molecular biology and cellular imaging that were unavailable, indeed almost inconceivable, eighty years ago.

In this theoretical framework attention once more turns from genes to the developing organism. DNA is no longer seen as an 'informational macromolecule' controlling the cell but rather as part of the web of molecules and their interactions that the cell employs during development. Information is thus generated by and during the developmental processes themselves. This is developmental systems theory, as proposed by philosopher Susan Oyama,[20] or autopoiesis, to use the terminology of cybernetician Humberto Maturana and biologist Francisco Varela.[21] Within this framework, at every moment of time an organism is both building upon its existing structures and generating new ones. Living creatures cease to be conceived of as passive vehicles, mere carriers for the all-important replicators, and are seen instead as self-organising and 'goal-seeking'. Biology's notion of goal-seeking, or teleonomy, is distinctly softer than the notion of teleology, with its sense of agency, of purposeful direction towards some final endpoint. Teleonomy refers to processes that give the appearance of purposefulness but are instead the result of physical and chemical mechanisms.

The other important implication of molecular epigenetics is the support it lends to the insistence that natural selection must work on entire lifecycles, not just on the adult. Darwin himself understood this well, but it was largely forgotten by many of his modern followers, fixated as they were on genes rather than organisms. Imagine for example 'a gene' which enables antelopes to run faster and escape predating lions. If the same gene also affects development such that the antelope matures more slowly, making it more vulnerable in infancy, then its potential benefit becomes a deficit. To rephrase Dobzhansky, 'nothing in evolution makes sense except in the light of development', now becoming reintegrated as evo-devo. As developmental biologist Anne McLaren (Foreign Secretary of the Royal Society and the first woman to be made an officer of that exclusive body) and close to the Waddington school, dryly observed of the rebranding, 'so we have really been doing evo-devo all these years'.[22]

EVOLUTION AND SELECTION

What are the implications of epigenetics/evo-devo for evolutionary theory? Selection can only operate if there are variant forms of a phenotype with differing fitness within a population. Such phenotypic variation can occur at many levels of organisation, from that of a particular DNA sequence, through its location within the genome as a whole, to the physiology, anatomy and behaviour of the organism. Further, with 20,000 genes and many trillion cells in the human body, it is evident that there is no necessary one-to-one correspondence between a change in a DNA sequence and a change in phenotype. Not only may a single change have multiple effects in many organ systems, because cells use DNA in multiple ways during development, but the change may also be entirely ignored. Cells are resilient ('robust' in Bateson and Gluckman's terminology) systems with many reserve mechanisms that can be brought into play to compensate for damage. Crick's Central Dogma of the one-way flow of information from DNA to RNA to protein is well and truly dead.

Changes in gene frequency thus do not necessarily produce a phenotypic change on which selection can act. Instead such changes in DNA may be hidden within the organism, until sufficient accumulate to produce a sudden phenotypic switch. This molecular mechanism underlies Niles Eldredge and Gould's observation that the fossil record shows millions of years of stability followed by periods of rapid evolutionary change. This is the theory of punctuated equilibrium – 'evolution by jerks' to its opponents – that has so enraged the orthodox evolutionary community wedded to Darwinian gradualism – 'evolution by creeps' to the punctuationists – and which led to charges that Gould was dragging Marxist revolutionary ideas into biology. His accusers choose to overlook the possibility that either Darwin had dragged Malthusian reactionary gloom into biology a century before, or that evolution by creeps could be read slightly more cheerfully as Fabian gradualism.

ORGANISMS AND ENVIRONMENTS

Within the media, and not infrequently when geneticists address the public, the organism and the environment are spoken of as two distinct unambiguous entities, always fixed and always knowable. For evolutionary theorists, particularly those who engage in wet biology and/or ethology, the situation is rather more complex.

Far from passively responding to a fixed environment, organisms – even those that on first sight may appear to be simple, such as bacteria – select and modify their environments. Put an E. coli bacterium into a glass of water and add a drop of sugar solution; the bug will swim towards the sugar, digest it, and move away to escape the waste materials that its digestion generates. Similarly, as Richard Lewontin points out,[23] which features of the world constitute a relevant environment depend on the organism. Bacteria are so small that they are constantly buffeted by the motion of the water molecules that surround them but are largely unaffected by the force of gravity. Water boatmen by contrast, indifferent to the motion of the water molecules, row along the surface of a pond supported by its surface tension.

Biologists understand organisms and the ecosystems in which they are embedded as evolving symbiotically. Consider the beaver dam, which begins with the arrival of beavers and a frenzy of dam building. Even as the dam is being built, it and the lake behind it become complex ecosystems full of many inhabitants and much interdependent activity. These mutual interactions mean that the dam, the lake and their inhabitants co-evolve through their generations. So too with humans and their symbionts – it is a chastening thought that there are more non-human – bacterial – cells present in the human body than there are human.

At the same time, radical environmental change in which entire species may perish can occur irrespective of how adapted individuals of the species were to the conditions before the catastrophe. If humanity were to perish, for example by our failure to respond to climate change, the urban rat population and the AIDS virus, which both depend on us, would disappear also. On the other hand, as Lynn Margulis points out, slime moulds would be the most likely organisms to remain and flourish. So much for our species supremacy!

A more profound rejection of genocentrism is made by Eva Jablonka and Marion Lamb in their book *Evolution in Four Dimensions*.[24] This draws on the extended debate about whether and how Lamarckian inheritance can be reformulated in a way that is compatible with modern biology. Can there be evolutionary change, under natural selection, independently of genes? Jablonka and Lamb argue that there are multiple levels at which selection occurs: genic, epigenetic, behavioural, and, for humans, symbolic. Human behavioural epigenetics is becoming a hot new research field.[25] Examples of epigenetic and behavioural selection abound in recent ethological research, from honey-bees and dung beetles to sticklebacks and snakes.[26] For instance, if pregnant rabbits are fed on a diet containing strongly flavoured foods, and continue to eat it while nursing their young, the young in their turn will prefer the same flavours, and in turn pass the preference on to their own young across several generations without the requirement for a change in genes. Given enough time, chance mutation may enable the genes to catch up with and consolidate such phenotypic change. The same is true for socially learned skills – from monkeys in Japan learning to wash potatoes before eating them to blue tits learning to open the aluminium foil tops of milk bottles delivered to British doorsteps in the 1950s to get at the cream. These socially learned skills too are transmitted across generations among the monkey and tit populations – but in the case of the tits were lost again when foil-topped milk deliveries became obsolete in the 1970s.

ADAPTATION, EXAPTATION AND CONTINGENCY

The most intense debates, however, continue to address the same questions that troubled Darwin, who emphasised that natural selection is not the only mechanism for evolutionary change. For him, sexual selection explained many seemingly non-adaptive features of the living world. Geographical separation – as of the famous finches on the different Galapagos islands – are another, as separated populations will gradually drift apart by chance or random variation quite apart from any other adaptive pressures. And what does one mean by adaptation anyhow? Ultra-Darwinians insist, for instance, that all the subtle variations that can be found in the banding patterns of snail shells are adaptive. Such rigid adaptationism famously came under attack by Gould and Lewontin at a meeting in 1979 commemorating the centenary of Darwin's death, at which they described it as a 'Panglossian paradigm'.[27] They pointed out that some seemingly functional aspects of an organism may be the accidental consequences of some quite other feature – as with, in their example, the spandrels (or spaces) between the arches which support the dome of Venice's San Marco. The spandrels are covered by glorious mosaics, which gives the illusion that they were designed to carry them, but architecturally they have no structural function. In Darwinian terms they are not adaptive. So too with biological features that might seem at first sight adaptive – such as the pattern of bands on a snail's shell – but are accidental consequences of the chemistry and physics of shell construction. Gould enraged many in his audience by devoting much of his allotted time to architectural rather than biological discourse, using it metaphorically to attack the determinism implicit in adaptationist theory. In the tea-break that followed his talk, Arthur Cain, a Liverpool evolutionary geneticist, denounced Gould as a Marxist, declaring that, as different shell colours camouflage the snails more or less effectively against predation by thrushes depending on whether the snails were in woodland or in open fields, it followed that every aspect of the patterning of the snail shells he studied had an adaptive function.

It was Gould and Elizabeth Vrba who gave the name to yet another potential contributor to evolutionary change: exaptation. An exaptation is some feature of an organism originally selected for one function that may also serve as the basis for another. The favourite example is that of feathers, believed to have evolved among small dinosaurs as a means of regulating body temperature, but which also enabled flight in the precursors of modern birds. For Gould such exaptations are an indicator of the chance or opportunistic nature of evolutionary processes, relying as they do on multiple mechanisms. If, as Gould repeatedly insisted, one could 'wind the tape of evolution back' to the pre-Cambrian or some other remote geological period and let it run forward again, it is very unlikely that sentient mammals like humans would appear.[28] That is, the evolutionary future is unpredictable.

This suggestion is unpalatable to those ultra-Darwinians who are happier with a more rigid determinism. The best adapted form – provided it conforms to physical and chemical laws – will always emerge, as Gould's most trenchant critic, Simon Conway Morris, has argued.[29] From the moment life emerged on earth some 3.5 billion years ago, it could have been predicted that something like conscious humans would in due course evolve – an argument strangely close to that of those who claim an 'anthropic principle' by which the entire universe is designed for human habitation, as the odds against it emerging by chance are too high. Such debates, while primarily focused on the past, have been extended into the future with the discussion of extraterrestrial life, intelligence and human contact.[30] Does natural selection predict that such extraterrestrials will be intelligent humanoids or is indeterminacy such that it can make no prediction as to the life-forms that will evolve?

FROM SOCIOBIOLOGY TO EVOLUTIONARY PSYCHOLOGY

Few of these debates within evolutionary theory have hindered the spread of the evolutionary metaphor far outside its biological domains, above all in the repeated attempts to at least tame and limit, and at worst to eradicate, the social in theorising humanity, and thus to biologise human nature. Support for the HGP drew on and in turn fostered the strongly hereditarian culture that had flowered during the 1980s. Genes were seen as inescapably determinist, not only for disease but for behaviour. Thus Mrs Thatcher claimed that there was no such thing as society – only individuals and their families, so that it was natural to wish to pass on wealth to one's children, while, as we mentioned in the Introduction, John Kennedy Jr. explained away his problem with alcohol as genetic. Such claims were sustained by a proliferation of research papers claiming to prove that everything from criminality to political and sexual orientation, racial prejudice and respect for royalty were inherited.

With the theatrics surrounding the announcement of the sequencing of the human genome, genes yielded to genomics, and DNA came to permeate speech and images from car advertisements to politics. Good design is apparently 'in the DNA' of the BMW, just as family values, according to David Cameron, are built into the

DNA of the Conservative Party. However, for philosopher Daniel Dennett and anthropologist Robin Dunbar,[31] to say that something is 'in the DNA' is rather more than BMW's metaphorical image of DNA sequences or the clichés of politicians or journalists. Evolution and DNA are integrated within a modern restatement of that same materialism which Darwin shared, in which the human mind itself is reduced to nothing but the epiphenomenal expression of a biochemical brain process – Moleschott's genius as phosphorus and Huxley's 'whistle to the steam train'. Having located humans within an anatomical and physiological evolutionary continuity, in *The Expression of the Emotions* and the *Descent* Darwin anchors 'mental powers' firmly to human biology. Human emotions and their expressions were for him the evolutionary descendants of those of our ape-like ancestors. Today it is not just emotion but the products of human culture, from art, music and ethical codes, that are argued to be manifestations of a gene-based process of natural selection, with their causal genes to be discovered by genomics and their phrenological locations demonstrated by neuroscience.

The debt of Darwin to Malthus is repaid as variants of evolutionary theory seek to renegotiate the boundaries between the biological sciences and the social sciences and humanities, so that what was once part of culture is increasingly naturalised – even, one might say, self-naturalised. We now have evolutionary ethics, evolutionary psychiatry and medicine, evolutionary aesthetics, evolutionary economics, evolutionary literary criticism, even evolutionary cosmology. This cultural colonisation is led by prominent standard-bearers. In *Sociobiology*, E.O. Wilson modestly proposed: 'It may be not too much to say that sociology and other social sciences as well as the humanities are the last branches of biology waiting to be included in the Modern Synthesis.' A few years later he less modestly proposed the idea of 'consilience', a unitary epistemology requiring nothing less than the subordination of the social sciences and humanities to the biological and physical. That handful of social scientists committed to sociobiology seem, unlike turkeys, to be voting for Christmas.

Just as Herbert Spencer sought the unification of scientific knowledge through the reduction of all the disciplines to one fundamental law, that of evolution, Daniel Dennett describes Darwinian natural selection as a 'universal acid' eating through every aspect of material and intellectual life in which less fit theories or artefacts are replaced by their fitter descendants.[32] The philosopher David Hull has argued that the history of theories in science can itself be seen as an evolutionary process powered by natural selection. The anthropologists Richerson and Boyd have employed the same argument to describe the changing design of Palaeolithic tools.[33]

Building on these premises and the Hamiltonian thesis of kin selection has been the mission of evolutionary psychology (EP), the most recent avatar of 1970s sociobiology. EP reverts to the Spencerian view of evolution as powered exclusively by competition. It ignores the many powerful arguments, suppressed in the heyday of the Modern Synthesis but being re-voiced today by many evolutionary biologists and philosophers, from Eliot Sober and David Sloan Wilson[34] to Martin Nowak,[35] that cooperation between members of a group or species, rather than just those individuals closely related genetically, and symbiotic relationships between species, can prove at least as strong a mechanism for selective change as does competition. (Very recently E.O. Wilson himself has made something of a U-turn, accepting the possibility of group selection, thus rejecting the rigid kin selectionism of his earlier books and earning himself a splenetic attack from Richard Dawkins as a result.[36])

While all biologists share the assumption that humans are an evolved species, evolutionary psychologists go much further, making the profoundly un-Darwinian assertion that 'human nature' was fixed in the Pleistocene and has not changed since. Their argument is that biological evolution has not been able to keep up with the rate of cultural change, summed up as 'stone age minds in the twenty-first century'. In making such speculative claims, which deny the inexorable processes of evolution, EP ignores evidence from both archaeology and population genetics. Archaeologists are well aware of how little we know about the social practices of early human societies from the fragmentary records of grave sites. The mediaevalist Tom Shippey, describing a seventh-century grave in

Yorkshire containing not only the remains of a rich young woman surrounded by treasures, but also on top of her coffin lid, another, older woman, pushed into the grave alive with her pelvis broken by stones thrown on top of her, asks whether this was a punishment burial or a human sacrifice? 'Archaeology cannot tell us about motives' he concludes.[37] And that's 150,000 years closer to us than the Pleistocene.

Furthermore, study of human genetic variance – the type of research programme that informed the HGDP and the IHMP – points to the speed with which changing agricultural practices, and hence behavioural epigenetics, have driven genetic change. For example, most human adults, like most other adult mammals, have difficulty digesting milk. Lactase, the enzyme present in infants that makes it possible to digest the milk sugar lactose, is inactivated as the child matures. However, over the past 3,000 years – a blink of an eye in evolutionary time – in societies that domesticated cattle, mutations that permitted lactose tolerance in adults spread. Today most adults in Western societies, as opposed to those in Asia, carry the mutation and milk is part of the normal adult diet.

ESCAPING BY SKYHOOK

Neither the body of evidence about the degree to which human physiology and anatomy has evolved in the thousand or so generations that separate us from our Pleistocene forebears, nor the fact that we have no idea of their psychology – and no way of knowing it from the artefacts and skulls and skeletons that remain – deters the armchair theorists. Consider the claims of evolutionary psychologist Marc Hauser in his book *Moral Minds*, revealingly subtitled 'How Nature Designed Our Universal Sense of Right and Wrong.'[38] It is not just that the requirements for living with one's conspecifics, or innate emotional responses to others' needs, may have helped shape moral codes. Rather, just as Chomsky argues that there is a universal linguistic grammar,* so for Hauser human-

* The paradox of Chomsky's position has often been noted: that of on the one hand holding a reductionist and highly conservative thesis about the origins of language, and, on the other, being a hugely influential social critic.

ity is endowed with a universal set of moral principles, independent of culture or social context. He acknowledges cultural variations (such as honour killing or homophobia) in how the principles are expressed, but argues that despite the variation there are underlying universals. However, if the expression of these principles is so varied, invoking an evolutionary imperative explains everything and therefore nothing. Hauser's political recommendations flowing from this imperative are disturbing in that he wants 'policy wonks' 'to listen more closely to our intuitions and write policy that effectively takes account of the moral voice of our species'. As it happens, Hauser wrote this before his own disgrace over data-manipulation was exposed, resulting in his high-profile papers being withdrawn and reluctant resignation from Harvard.[39] It seems that his thesis of universal morality has thus to be weighed against the influence of competitiveness and a desire for fame. Hauser's thesis on the neuroscience of ethics ran afoul of the ethics of neuroscience. Perhaps this is why in *Moral Minds* he hedged his bets by suggesting that the policy wonks should not blindly accept universal morality since some of our evolved intuitions are 'no longer applicable to current societal problems'. An all-encompassing theory with a get-out clause as big as this scarcely rates consideration.

There is a further puzzle: for no sociobiology or EP theorist, from cognitive psychologist Steven Pinker to Wilson, Hauser and Dawkins, is it apparently obligatory to obey the demands of our selfish genes. The closing sentences of Dawkins's *The Selfish Gene* explain that 'we' humans, albeit only 'lumbering robots', can, unlike other species, escape their tyranny. For Wilson, a less sexist society can be achieved if 'we' wish, though at the cost of a loss of 'efficiency'.[40] For Pinker 'evolutionary explanations of the traditional [*sic*] division of labor by sex do not imply that it is unchangeable, or "natural" in the sense of good, or something that should be forced on individual women or men who don't want it.'[41]

When Pinker tells us that he has decided not to have children, or that he can tell his genes 'to go jump in the lake', or when Dawkins escapes the tyranny of his selfish genes, by what process do they deny this genetic imperative? Is there a location within the brain, a gene for free will? The theorists of mind and genetic determinism are silent. Their sense of personal agency is everywhere evident, but their theory provides no explanation, they escape by a rhetorical skyhook. Yet given that, like Pinker et al., we understand ourselves as thinking, moral, emotional and deciding beings, the issue of human agency is not a matter that can be ignored. The combination of genetic determinism and escape by skyhook lies at the heart of a neoliberal culture: the affirmation of individualism.

'GREAT IS OUR SIN'?

Like that of the HGP itself, Darwin's legacy is thus ambiguous. This century's ongoing reintegration of evolution and development – which Darwin would surely have welcomed – has built upon the discoveries of genetic complexity revealed by the sequencers while at the same time demonstrating the limitations of gene talk. Human life, pace Walter Gilbert, is not to be read off from a string of letters inscribed on a CD. But the underpinning social assumptions built into Darwin's original formulations – of competition, of the inevitability of class, race and gender inequalities – have had a malign effect on the subsequent evolution of evolutionary theory itself. Sociobiology's and evolutionary psychology's project, as Edward Wilson so unambiguously proclaimed, is to render the social sciences consilient with – read: subordinate to – the life sciences. But even Darwin himself had moments of doubt. In *The Voyage of the Beagle* he observed that 'if the misery of our poor be caused not by the laws of nature, but by our institutions, great is our sin.'[42] In the context of the crisis of the political economy of globalisation, his reflection is as salient today as when it was written.

3

Animals First: Ethics Enters the Laboratory

ANIMALS FIRST: A VERY ENGLISH STORY

Darwin is widely seen as a kindly liberal man but one who did not engage in politics: reading, researching, publishing, endless letter writing and spending time with his family and friends kept him fully occupied after his *Beagle* years. Nonetheless, in the 1870s he took an active part in the first ethical and political conflict between 'science' and 'society'. The issue at stake was whether the deliberate infliction of pain and suffering on animals was intrinsic to biological research. Opposing the new experimental science of animal physiology, the increasingly powerful antivivisectionist movement was planning to introduce legislation to prohibit all animal experiments. Darwin, his fellow scientific researchers, and the new ambitious doctors committed to scientific medicine (think of *Middlemarch* and the young Dr Lydgate's ultimately failed project of advancing scientific medicine) mobilised themselves to defend physiology.

The antivivisectionist movement had two main strands: the hunting, shooting and fishing aristocrats and landed gentry; and the social and political feminist movement, predominantly composed of upper-middle-class women. The two were not obvious allies. The former were part of a pre-capitalist formation and thus no supporters of the science or the scientists who were integral to the modernity eroding their world; while the latter, in their pursuit of equality between the sexes, were potential allies of modernity. Capitalist political economy has long been patriarchal, but it is not necessarily so, at least for highly educated women. Feminism, like most social movements, was not without its divisions. While for most feminists vivisection was evil and news of the struggle was reported regularly in the leading feminist journal, *English Woman's Review*, an influential minority – including the pioneering doctor Elizabeth Garret Anderson and the novelist George Eliot – welcomed biomedical science and recognised the case for animal experimentation.

The campaign against cruelty to animals was an established part of the English cultural landscape, with the dog, 'man's best friend', and the noble horse as central. Anna Sewell's 1877 novel *Black Beauty*, over which generations of girls wept, was a classic of the genre, mobilising its readers' sympathies in defence of the horse. While horses, unsuitable as experimental animals, were relatively safe from the vivisectors, dogs were not. Size, cost and availability made them obvious subjects. Dogs have indeed been the animals closest to us, not in any evolutionary sense, but as the first animals to live alongside, and then with, humans. Fossil and archaeological evidence indicates that dogs hung around the small groups of pre-historic people, scavenging human faeces as an easily available food supply. If not our best friend, they are certainly our oldest friend, and they retain their taste for the malodorous. The family dog is just that – part of the family. To borrow from Orwell, when it comes to animal protection both within and without the laboratory, all animals are equal but some are more equal than others. In the twenty-first century better to be furry with big eyes than crunchy or slimy.

Even though opposition to cruelty to animals was thus something of a cultural given, just what constituted that cruelty was not obvious. Queen Victoria, patron of the Royal Society for the Prevention of Cruelty to Animals, and her fellow aristocrats saw no cruelty in their passion for blood sports (George Bernard Shaw was later to call them the 'fur and feather brigade'), insisting rather on the higher moral ground they occupied, as against the calculated cruelty of the parvenu experimentalists. Feminists denounced the physiologists as the 'new priesthood', seeing a parallel between the male experimenters deliberately inflicting pain on animals and husbands beating their wives, or the degradation of medically managed birth. One suggestion, no weirder than many others made at the time, was that Jack the Ripper was perhaps a physiologist 'delirious with cruelty'. In return, the experimental physiologists denounced the feminists as the 'shrieking sisterhood'.

A century later, at a Cambridge meeting discussing the birth of laboratory medicine, Hilary, like the other feminists present, was outraged when one male historian of science interjected: 'You can do anything you like in a laboratory or a whorehouse.' He apologised but the small conference group was acutely embarrassed. Yet the Freudian slip which compared the 'you' in the laboratory to the 'you' in the whorehouse was all too gendered. At the level of the masculine unconscious do men still understand themselves as licensed to be wantonly cruel to women and laboratory animals alike? While the state has moved with some efficiency to regulate the welfare of animals, the continuing high levels of violence towards women in the sex trade and in the home gives no comfort.

If the feminist opposition to vivisection was highly organised, working-class opposition was more a matter of smouldering distrust, with occasional ritualised conflicts typified by the fight between medical students from University College London (birthplace of animal experimentation) and working-class young men for possession of the statue of a little brown dog erected by the antivivisectionists in Battersea Park. The distrust lay in the origins of the public hospital as an institution for the free medical treatment of the poor, with the unstated quid pro quo that the doctors were entitled to experiment on their patient-subjects in life, and remove body parts or whole skeletons after death, as resources for research and teaching.

Darwin, the great naturalist, was also a sophisticated experimentalist, though his chosen subjects, earthworms and barnacles, fell well below the radar of the English hatred of cruelty to animals. He had earlier abandoned his medical studies in Edinburgh, repelled by the violence of surgery without anaesthesia. He also exchanged correspondence with women botanists and had met the feminist and freethinker Frances Power Cobbe, with whom he had discussed his theory of evolution. She thought well of him and his kindness and wrote that 'he would not allow a fly to bite a pony's neck'.[1] Cobbe was to become a leading figure in the antivivisection movement as well as in the campaign against wife-beating. But her kind Mr Darwin was committed to the new experimental physiology – even though, as the physiologists themselves acknowledged, their research required the 'method of pain'.

The method of pain meant experimentation on animals, usually dogs, sometimes without anaesthesia, treated with curare to paralyse them but still leaving them able to feel pain. The method had been pioneered by the physiologist Claude Bernard in France, and followed in Britain by John Burdon-Sanderson in the 1860s. For Bernard, the new 'science of life' was 'a superb and dazzlingly lighted hall which may be reached only by passing through a long and ghastly kitchen'.[2] For him, vivisection served to elucidate the processes of digestion; for Burdon-Sanderson the workings of the heart. Their shared vision was of a laboratory revolution that would do nothing less than put the art of medicine onto a sound scientific footing for a new generation of scientifically trained doctors. These would learn by doing hands-on experimentation as part of their training and would transform the practice of medicine. From within the Cartesian world-view that dominated in France, in which an animal's cry of pain was merely the squeak of rusty machinery, Bernard could set aside the pain and suffering of the unanaesthetised animals on which he worked. However, given that Bernard's wife and daughters were horrified by his experiments, Burdon-Sanderson's widow left her estate to the protection of animals, and that Darwin did not want the Down house ladies to overhear the plans for defence, it seems that the male experimenters were aware that they would not be able to convince even the women in their families that the cruelty was justified.

In 1865 Bernard made two powerful related claims in his *Introduction to the Study of Experimental Medicine*. Ethically, he argued that except in rare cases research on animals should precede experiments carried out on human patients. Epistemologically, he proposed that animal research permits the scientific study of the normal while research on human patient subjects only tells us about the abnormal. The former were thus able to contribute to scientific knowledge in the way that the latter could not. The model Bernard proposed for the translation of research from bench to bedside became the norm, and importantly served to protect the human subject from premature experimentation.

For Burdon-Sanderson, located in the trickier milieu of England, experimental research had become ethically and politically more problematic. The combination of the two wings of the antivivisection movement was powerful. The laboratory revolutionaries were indeed right in their assessment of the threat, and by 1875 the antivivisectionists were powerful enough to introduce a bill in the House of Lords to ban animal experimentation. The government, under pressure from the scientists, responded with the classic British technique of establishing a Royal Commission. The physiologists defended their research to the Commission on the grounds that although they inflicted pain, this was not motivated by cruelty but by the search for scientific understanding, and therefore would benefit medical care. Such was the strength of public feeling, there was a real danger that animal experimentation, and with it the future of life science, would come to a halt in Britain if nowhere else. Determined to prevent this, the researchers began mobilising by establishing a professional society (the Physiology Society), enlisting Darwin in support. He responded by sending his recently graduated medical doctor son, Francis, to join their discussions. The compromise proposed by Darwin and his colleagues, and accepted by the Commission, was that animal experimentation should be regulated so as to minimise cruelty by ensuring that experiments be performed under licence and wherever practicable under anaesthesia. The 1876 Act established an inspectorate under the Home Office, but since the first inspectors were themselves physiologists and Fellows of the Royal Society, their inspections were unlikely to be too demanding. The Act did not satisfy the antivivisectionists, who in 1898 established the British Union for the Abolition of Vivisection; the physiologists responded by setting up a Research Defence Society. Both are still extant today, but the centre of concern about animals is now the Animal Rights Movement.

Over the last century the cultural understanding of both experimentalists and the public as to which animals are closest to us has changed. Nature films have played their part. Who can forget the image of David Attenborough being cuddled by a huge mountain gorilla? Evolutionary theory, aided and abetted by television documentaries of animals in the wild, has triumphed. It is no longer man's best friend that demands protection but the primates, our closest relatives. This change is strongly reflected in European legislation which bans research on the great apes, and although it still permits limited research on monkeys, they must be laboratory bred, not taken from the wild. Opposition over the past decade to two proposed institutes for primate research led to the Cambridge project being cancelled – although the Oxford one faced down the protesters, helped in good part by student supporters. Their presence was politically significant as today's activist students are more likely to be found on the other side of the barricades.

CHANGING THE SUBJECT

The origins of bioethics have been read, broadly speaking, in two main ways. The first explores the history of ethical thinking, through the world's great religions and classical thought to modern Western philosophy, and the guidance these offer to those who profess medicine. While religious influences are still present, the interest of the patient has long been seen within the West as sufficiently protected by the Hippocratic Oath of doing no harm. Interpretation, barring manifestly criminal acts on the part of the doctor, rested on the prevailing consensus of the profession together with the religious commitments of the individual doctor. As scientific research on both patients and healthy subjects became central to the practice of modern medicine, reliance on the Oath, essentially a form of professional and personal self-regulation, as providing sufficient protection for the patient has increasingly been seen as inadequate to prevent abuse, whether in the clinic or the laboratory.

A new ethical code, and even external regulation, was required. Formalising this took the best part of the twentieth century. New terms have been coined: medical ethics, concerned with the doctor–patient relationship; bioethics, relating to research on both patients and non-patients; and biomedical ethics, covering both care and research. But despite attempts by their standard-bearers to define these as unequivocally separate activities the boundaries have always been porous. Here we mostly use bioethics to refer to the ethics of biomedical research on humans.

The second approach to reading the history of bioethics, which we share, places much less emphasis on the history of ideas and more on historically situated accounts of particular research scandals coming into public view and thereby precipitating outrage and with it the demand for professional, even legislative, corrective action. Typically it has been the revelation of the lack of ethical regard in public health research carried out on human subjects with little or no social power to give or refuse cooperation – whether prisoners, soldiers, learning disabled people or other incarcerated people. In practice, these ethical responses, while often invoking the language of a universal ethic, draw on the prevailing locally dominant theologies and philosophies – not excluding Tom Paine's common sense.

For the English, ethics entered the laboratory with animals, but for the Germans the ethical concern was with human research subjects as presenting a new problem not covered by the Hippocratic Oath. The story begins with a scandal from 1896, when the public learnt that serum from a patient with syphilis had been injected into patients without syphilis. The ensuing furore led the physician Albert Moll to propose a moral contract between the doctor and the patient as research subject to prevent any repetition of such abuse. The term bioethics was first introduced in 1927 by the German protestant pastor Fritz Jahr, who recognised that the growth of science and technology demanded a new ethical framework. He argued that Kant's moral imperative of treating others as ends not means should be taken as a bioethical imperative to be applied to all living forms – that is, bios.

Bioethical thinking became practice following a scandal of public health research. A study on the effectiveness of a tuberculosis vaccine carried out in an institution for learning disabled children had caused the death of seventy-five. The Hippocratic Oath of doing no harm had failed to protect them. The public outcry that greeted the news of the children's deaths led the Weimar Ministry of the Interior to issue a directive in 1931 requiring the voluntary consent of research subjects. This was essentially a political/ethical response to a public moral scandal. But by then the end of the Weimar Republic was in sight and the election of the Nazi Party imminent.

THE DOCTORS' TRIAL AND THE NUREMBERG CODE

The war crimes trial of the Nazi leaders by a four-nation International Military Tribunal began in Nuremberg in October 1946. As there was no legal precedent for such a trial, the political decision to prosecute was taken by the four allies (the British, Americans, French and the Soviet Union) at a meeting in London in 1945. Although sceptics suggest that the trials were simply an example of 'victor's justice', and hence a less than pioneering phenomenon, their inevitability has been questioned, not without irony, by the High Court Judge Stephen Sedley, who noted that although Churchill was all for shooting the Nazis out of hand, President Truman was committed to an international trial. Stalin, who certainly understood the value of a show trial, shared the American enthusiasm for a very public event. The spectacle of the formal process of prosecution and defence being played out before the world's press would mean that Nazi crimes would receive maximum publicity and thus ensure that international opinion was hardened against them. Additionally the fate of the convicted would serve as a deterrent against further such crimes.

The first and most important trial was that of the major German war criminals. Of the twelve that followed the first was the 'Doctor's Trial', begun in December 1946 and concluded in the autumn of 1947. Preparations involved a British, American and French International Scientific Commission drawing on material collected by expert groups who had followed the advancing allied armies in search of evidence of what would now be called abuse of human rights (and less publicly valuable military research) during the final years of the war. The trial itself was conducted by an American military tribunal, with US judges and lawyers (both for the prosecution and the defence). Its conduct was, on the surface, entirely scrupulous. The defendants by no means denounced themselves by confessing to their crimes, as in the classic Stalin show trial; instead they remained defiant. They defended their research ethics, arguing that their practices were no different in principle from those carried out by biomedical researchers in the US and other countries. Like them they were doing 'good science'.

Because the trial focused on war crimes, Nazi crimes against German citizens prior to the outbreak of war were excluded. This was no minor exclusion, as no fewer than 400,000 'mentally unfit' German citizens were murdered during this period. Doctors were central to Nazi eugenics. Unsurprisingly, as the key group proposing, planning and implementing racial hygiene, medicine had the highest proportion of Nazi members of any profession. Reducing the numbers of the unfit was a public service, not a crime. As we will explore in the next chapter, the question of the 'mentally unfit' and how their presence weakened the national stock haunted the intellectual and political classes of the West. Since they shared with the Nazis the eugenic project of reducing the numbers of the unfit, prosecuting the murderers would not only have extended the charges beyond war crimes but potentially opened up the uncomfortable political question of what other nation states had done and were doing to their 'mentally unfit' fellow citizens. Instead, the Nuremberg prosecutors focused on wartime experiments conducted on foreign nationals: Jews, Roma, communists, socialists or the physically or mentally 'unfit'. This was seen as a means of avoiding an unwelcome precedent for intervention in the internal affairs of a state (today such illegal intervention is no longer a matter of international consensus, the grossest example being the 'humanitarian intervention' in Iraq, with the mission creep in Libya making it a close second).

Twenty-three biomedical researchers, twenty of whom were physicians, were accused of murder and torture in the conduct of medical experiments on concentration camp inmates. Their experiments studied the effects of exposure to freezing, low oxygen pressure, wound healing and infection – all questions of potentially equal military interest to the victors. Sixteen defendants were found guilty, of whom seven were hanged, nine sentenced to various terms of imprisonment, and seven acquitted. Among the acquitted, one, Hubert Strughold, an active Nazi involved in the freezing experiments, was singled out for protection by the US Joint Chiefs of Staff and recruited into US military biomedical research. Strughold was but one of a group of such scientists, often with a Nazi background, recruited by the US from post-war Germany in what was called Operation Paperclip.[3] Similarly Japanese microbiologists from the notorious Unit 731, which had been responsible for horrendous experiments with biological weapons, first infecting hundreds of thousands of Chinese civilians of all ages, including infants, then vivisecting them without anaesthesia during their occupation in the 1940s were not tried like the Nazi doctors but were recruited into the US biological warfare research establishment at Fort Detrick in Maryland; it was not until many years later that their crimes became public knowledge in the West. (Some documents were released by the US government only in 2000.)

Denazification post-1945 was conspicuously uneven on both sides of the Iron Curtain, certainly so far as top scientists were concerned. Valued abroad by the wartime allies, at home they soon regained their scientific respectability. In West Germany the leading theorists of scientific racism, including some who had participated in the camp experiments, not only came out unscathed but were restored to their academic positions. Only when the German geneticist Benno Muller-Hill's *Murderous Science* documented the phoenix-like survival of the Nazi doctors and scientists, those theorists and practitioners of scientific racism who had defined both the Aryan superman and the *untermenschen*, did the limits of denazification become apparent.[4] The key figures of Nazi race science did not merely survive, they became directors of leading laboratories, welcome participants in the scientific community.*

At the trial itself, the lead prosecution lawyer, Brigadier General Telford Taylor, began with the observation that although the charges against the doctors were inescapably about murder, the Nazi doctors' research had involved 'more than' murder. The legal challenge for the prosecution team was to provide an overwhelming case to the court which defined precisely what this 'more than' was. Was it what Taylor referred to as 'thanatology', the science of producing death, or was it the scientists' total indifference to the suffering of their research subjects, or both? Because of the commitment to legal process the trial was not simply a pushover. The prosecutors were confronted by the defendants' self-confidence, even in the box and charged with heinous crimes, stemming not only from their convinced Nazism, but also from their secure self-identity as biomedical research scientists. While not a position their victims could be expected to share, the sufferings of the research subjects could be, and were, entirely discounted within an ideology of scientific racism which insisted that they were not human beings, and were therefore outside the purview of biomedical research ethics. Some 1,000 Russians, 500 Poles, 200 Jews and 50 Roma were thus defined as having the moral status of laboratory mice,[5] but with the undiscussed advantage of having human bodies. The ethics of using knowledge gained in the 'ghastly kitchen' of the Auschwitz experiments remains contentious; for some, knowledge, like money, has no smell; others are less sure.

The Nuremberg defence team challenged the prosecution's argument that the Nazi's research entailed 'more than' murder by pointing to a large-scale US malaria study in Illinois in the 1930s, which involved prisoners as subjects, albeit as 'volunteers'. The defence argued that the malaria study was not fundamentally

* When Steven, at a neuroscience conference in Europe in the 1980s, objected to work by Mengele's collaborator Verschaur being cited as if it were part of normal science, he was criticised by the researchers present for embarrassing the meeting by 'bringing politics into science'. To the embarrassment of its director, neuroscientist Wolf Singer, tissue sections from camp victims were discovered in the museum of a major brain research institute in Frankfurt in the 1980s.

different from the Nazi 'experiments' in terms of the freedom from coercion possible for the US prisoners, however subtly deployed. The choice of Andrew Ivy, who was based in Illinois and had supported the malaria study, as the prosecution's key research ethics adviser and witness, gave the defence team an opening to question the prevailing ethics of biomedical research practice in the US, and Ivy's relationship to the malaria studies in particular.

What the Nuremberg prosecutors needed was to be able to point to a precursor ethical code, and Ivy was crucial in providing this, appearing as a witness for the prosecution along with the American neuropsychiatrist Leo Alexander and the German medical historian Werner Leibbrand. While Alexander was as committed as Ivy to the production of a code of ethics, Leibbrand had a rather different perspective. He was a German doctor who had lost his position because of his wife's Jewish ancestry. He had become a medical historian, surviving precariously in Germany and somehow outliving the Third Reich. Called by the prosecution at the trial, he argued that since the beginning of the twentieth century, German medicine had come to regard the patient as 'a mere object, like a mail package', not so much a person as a 'series of biologic events'.[6] His analysis of the dehumanisation and objectification of patients was thus much more radical than the seeming avant-garde ethical debates of Weimar Germany. He saw little distinction between Nazi and US biomedical research, in that neither had secured their subjects' free and informed consent. Despite Leibbrand appearing for the prosecution, his courageous evidence was thus highly inconvenient to them.

Ivy was a medical physiologist and an influential figure within the American Medical Association. He argued that there was a practice of bioethics in the US even though it was not formally codified, a position he argued more with the confidence of a medical politician than with evidence. Medical ethics governing the relationship of the clinician to the patient were indeed in place, but biomedical research ethics were not. To provide the 'more than' that Taylor needed, Ivy and Alexander drafted such a code and ensured that the American Medical Association endorsed it in 1946, immediately prior to the trial. Ivy's draft, along with that of Alexander and others, was finally merged in the Nuremberg Code published as part of the judges' summing up at the end of the trial. Evelyne Shuster, a historian of the trial, concludes bleakly that 'the primary objective of Ivy's medical ethics principles was to ensure that human experiments were possible in the future. All other issues, like the protection of human and patient rights in medical science, or the role of the informed consent principle were secondary to this overarching objective.'[7] There is more than an echo of Darwin's and his colleagues' legislative proposals; for the scientists of both the nineteenth and the twentieth century the crucial issue was to ensure the future of biological experimentation, even if some compromise was necessary. But now the research subjects were not animals but humans.

Neither Alexander nor Ivy saw the Nuremberg Code as needing to be applied to US research. The monstrosity of Mengele and his colleagues' research could not possibly happen in the US. In consequence, after Nuremberg, each reverted to his pre-war physician-centred ethics. Alexander thought that the Hippocratic view of research ethics coincided with the intent and vision of the Nuremberg Code, and he did not distinguish his research from the ethical obligations of his profession as a doctor. Ivy, despite his carefully timed encouragement of the AMA to adopt a precursor to the Code, seems to have seen no need to ensure that US biomedical research was governed by ethical considerations. For most it was business as usual; research carried out in a democratic country was by definition ethical. Ironically, Ivy's own career ended in ethical disgrace, as he became embroiled in controversy over a quack cancer cure. He attempted to block well-designed control trials of the efficacy of a drug, Krebiozen, for whose manufacturing and marketing he had served as a consultant.[8] It is not difficult to see why those concerned by the issue of victor's justice are troubled by the roles of Ivy and Alexander in the trial and the setting aside of Leibbrand's evidence.

Crucially the Nuremberg Code moves beyond the reliance on protection by professional ethic – the paternalistic protection offered by the Hippocratic oath – and insists on the centrality of the moral agency of the patient-subject in biomedical research. Its key requirements for securing this new status were that the patient-subject must be informed of the purposes of the experiment, must be free to give or withhold their consent, and must have the unqualified right to withdraw from the research. Research on animals must precede that on humans. Furthermore, the research should be for the good of society and be such that it could not be carried out in any other way.

These were inspirational principles, but those who formulated them gave little or no attention to their implementation. Unlike the United Nations Charter of Human Rights the Code was not disseminated even in the US. Although Jonathan Moreno, one of the more acute historians of the abuse of research ethics,[9] observes that Nuremberg had little impact, he fails to ask just how biomedical researchers were supposed to know about this new moral agency of their subjects, and what they were supposed to do to bring ethics into their research. Nor was any thought given to the matter of enforcement against any future breach of the Code – not least if the breach occurred in a country with little or no public opposition. Would-be enforcers of the Code found themselves as mice faced with the cat-belling dilemma.

THE WORLD MEDICAL ASSOCIATION AND THE HELSINKI DECLARATION

Given the horror of the Nazi research revealed at the Doctors' Trial, it was crucial for biomedicine to turn the Code's ethical vision into professional ethics. This became one of the objectives set by the new World Medical Association (WMA). In July 1945 an informal conference of doctors from several countries was held in London to plan the setting up of an international medical organisation. Two preliminary meetings in 1945 and 1946 were held, followed by the first international congress in 1947 with twenty-seven affiliated national associations. Their agenda was formidable, ranging from medical ethics, professional education and socio-medical issues to the ethics of biomedical research. The first formal Declaration of the

WMA, made in Geneva in 1948, reasserted physicians' responsibility to their patients, but it was not until six years later, at its meeting in Helsinki, that the WMA formally addressed the issue of research ethics. The ensuing Helsinki Declaration was in many respects a restatement of the ten principles of the Nuremberg Code, but with the improvement of distinguishing clearly between patient-subjects and healthy experimental subjects. However, while for Nuremberg voluntary informed consent was the first and overriding principle, in the Helsinki Declaration it is relegated to ninth out of twelve 'basic principles':

Helsinki's Principle 9 states:

> In any research on human beings, each potential subject must be adequately informed of the aims, methods, anticipated benefits and potential hazards of the study and the discomfort it may entail. He or she should be informed that he or she is at liberty to abstain from participation in the study and that he or she is free to withdraw his or her consent to participation at any time. The physician should then obtain the subject's freely given informed consent.

Principle 2, however, introduces the concept of good science as a key criterion:

> The design and performance of each experimental procedure involving human subjects should be clearly formulated in an experimental protocol which should be transmitted to a specially appointed independent committee for consideration, comment and guidance.

The centrality of peer review with its criterion of 'good science' in Principle 2 was seen by those clinicians for whom human rights were the core value of the Nuremberg Code as dangerously opening the door to the Nazi doctors' claim that they had been doing good scientific research. Did Helsinki unwittingly rehabilitate the Nazi doctors' prioritisation of 'good science' over human rights? Did it weaken the protections afforded to research subjects? The problem does not, and perhaps cannot, go away.

Since 1964 there have been a series of revisions of the original

Declaration, the most recent being in 2008, each one increasing its length and coverage. These have strengthened the emphasis on peer review for research on human subjects via Institutional Review Boards, required greater ethical justification for research projects, formalised the legitimacy of the use of placebos in double blind experiments, and, especially in the aftermath of the AIDS crisis, attempted to develop standards for trials in developing countries. Post Helsinki, national codes began to be developed, initially self-regulating and very slowly brought under regulation with – in most but not all countries – enforceable penalties for breaches.

The history of the World Medical Association has not been without its ethical and political scandals. The election of the sometime Nazi Hans Joachim Sewering to the WMA presidency in 1992 was the most conspicuous. Sewering had joined the Nazi Party in 1934, and worked as assistant physician at the Schoenbrunn Sanitarium. According to two nuns who broke their silence in 1993, he was responsible for the death of 900 physically and mentally disabled children by transferring them from Schoenbrunn to the Eglfing-Haar 'healing centre', a euthanasia facility south of Munich. He was specifically charged over the death of fourteen-year-old Babette Fröwis, who was to become symbolic of the victims of Nazi euthanasia. Diagnosing Babette as epileptic without even seeing her, Sewering sent her to Eglfing-Harr. He flatly denied these allegations and after the war became a respected doctor, and the head of the German Medical Association in 1993. However, the evidence supporting the allegations against him was sufficiently well founded that in 2005 the US was prepared to prosecute him over war crimes.[10] Despite this, in May 2008, when Sewering was in his nineties and still living in Dachau (an irony that seems to have escaped the prize-givers), the German Federation of Internal Medicine awarded him the Gunther-Budelmann medal for 'services to the nation's health system', its highest honour. Although international protest ensured that Sewering was forced to resign before he could take up office as WMA President, US Physicians for Human Rights asked how a former Nazi could have been active within the WMA for over twenty years without being exposed.

Putting the principles of the Code and the Declaration into practice has been a slow process, particularly in public health and military research. The latter has been structurally invisible, concealed by national security, but, like the hidden part of an iceberg, is often the largest part of the national research system. (In Britain since 1945, defence research has year on year taken between 30 and 60 per cent of the country's research budget.) Military research routinely used soldiers in no position to give or withhold their 'voluntary consent'. Nor were the unsuspecting individual civilians and CIA staff given LSD as part of the CIA experiment known as MK-ULTRA offered any options.[11] From the 1940s to the 1970s researchers at the US Atomic Energy Commission (AEC) had exposed Alaskan villagers to radioactive iodine, fed forty-nine learning disabled and institutionalised teenagers radioactive iron and calcium in their cereal, exposed 800 pregnant women, psychiatric patients and prisoners in San Quentin to radioactive material, and injected seven newborns with radioactive iodine.[12] Field trials to test the dispersal of biological weapons were carried out by releasing bacteria into the New York subway system, and spraying from planes or ships off the Californian coast. Many of these experiments and trials – notably those involving potential 'mind-bending' drugs and procedures – recruited distinguished academic scientists, including Donald Hebb, a central figure in the pioneering years of neuropsychology.

Thus for decades post Nuremberg and Helsinki, and in the pursuit of such militarily inspired goals, Codes and Declarations were set aside by organisations such as the US Defence Advanced Research Project Agency (DARPA). Comparable practices were of course not confined to the US, where freedom of information legislation has allowed much to be brought to light.[13] Less is publicly known about equivalent research undertaken by other major military powers. In Britain, where Freedom of Information (FoI) requests have only recently become possible and are met with routine obfuscation (and in the case of the NHS risk register in 2012, direct refusal) complaints by ex-soldiers revealed that during the 1950s at Porton Down, the site of the UK's Chemical Defence Experimental Establishment, military personnel were the routine subjects of experiments with nerve gases without being told the nature of the substance or the potentially severe risks entailed (some were deliberately misinformed, being told they were taking part in trials of cold vaccines). Some died, others were severely brain damaged, but it took years of pressure and legal argument before they received any redress. Indeed, only in January 2008 did the UK MoD finally propose a settlement with the remaining survivors of experiments conducted half a century previously – of a mere £8,300, rather less than six-months pay at the 2008 average level.[14]

There was little or no ethical firewall between military research and patient care. In 1966 the Director of Porton's Microbiological Research Establishment (MRE), C.E. Gordon Smith, together with colleagues at St Thomas' Hospital in London, reported the experimental treatment of twenty-eight severely ill leukaemia and other cancer patients with langat and kyasanur viruses on the speculative grounds that these viruses caused the death of white blood cells and might therefore mitigate the cancers. These viruses were also of interest to MRE as potential biological warfare agents. Most of those treated died, some within hours of the injections; four improved transiently. While desperate medical situations can and do demand desperate measures, this research challenged the belief that the Hippocratic Oath bidding doctors to do no harm was adequate. Military interest trumped patient protection.[15] Tests on the dispersal of bacteria via the London underground system analogous to those in the US were also conducted secretly. But the most egregious cases were those involving military personnel and the local Aborigine people exposed to radiation during the atomic and hydrogen bomb tests in the Pacific and on the Woomera range in Australia between 1955 and 1963. Although the surviving veterans have belatedly received some compensation, the chances of the Aborigines – even in the twenty-first century and with the support of human rights lawyers – receiving recompense via the courts are zero. Like the veterans they seek justice; unlike the veterans they have few resources and little political clout.

TUSKEGEE AND THE CIVIL RIGHTS MOVEMENT

US bioethics were born not out of philosophical or theological reflection on Nuremberg but out of the ferment of social unrest of the 1960s with the rise of the new social movements. It was in this context that the civil rights movement, and indeed the whole of America, came to learn of the Tuskegee syphilis study carried out by the Public Health Service (PHS). The purpose of the study, begun in 1932, was to map the natural course of the disease among poor black Americans at a time when syphilis was common among the black Alabama population. Of the 600 in the study, some 400 already had syphilis. The subjects were told that taking part would entitle them to health care, even though the PHS had no intention of providing any treatment and indeed actively stopped other health-care agencies from offering treatment. The men were neither told they had syphilis nor offered any drugs, even though Salvarsan 606, a risky mercury-based drug, had been available since 1910. When penicillin, a highly effective drug against syphilis, became available in the 1950s, the PHS began using it to treat other patients, but not the Tuskegee men.

The persistent efforts of a whistle-blower from within the study were finally successful when in 1972 the outrage of Tuskegee finally appeared nationwide as a front-page story. Given the race riots that were engulfing American cities, delay was not an option, as Senator Edward Kennedy, already involved in the civil rights campaign, immediately saw. He established Congressional subcommittee meetings in early 1973. These considered the Tuskegee case in April, and in the same month the Senate formally criticised the Federal government for failing to act more swiftly. In the same year, a class action suit was filed in a US District Court on behalf of the study subjects. In December 1974, the government paid $10 million in an out of court settlement. Compensation at this speed and at this size could only have happened in the context of the threat to civil order posed by a profoundly alienated and angry black America.

The extent of the racist ethical abuse was graphically revealed in an interview given by John Heller, director of the public health venereal diseases unit from 1943 to 1948, to James Jones, the historian of Tuskegee. Among Heller's responses was the following: 'The men's status did not warrant ethical debate. They were subjects, not patients; clinical material, not sick people.'[16] What is the difference between the research ethics of the Nazi doctors and those of this public health researcher? Or those of distinguished virologist Chester Southam at the Sloan-Kettering Institute for cancer research, who in 1954 injected cancer cells into the arms of hospitalised patients and prisoners to test the hypothesis that cancer might be infectious? Only when three young Jewish doctors at the hospital resigned rather than endorse his experiment, citing Nuremberg, did the case come to light. Southam's wrist was lightly slapped, though it did not prevent him from later becoming President of the American Association for Cancer Research.[17] Around this time sociologist Bernard Barber and colleagues carried out a pioneering study of ethical control in medical experimentation, concluding that there was little or none.[18]

In 1970, neuroscience research showing that the amygdala, a deep region of the brain, was associated with emotion lay behind a proposal of neurosurgeon Vernon Mark and psychiatrist Frank Ervin. They proposed to 'cure' ghetto rioters and revolutionary black prisoners by removing the offending brain region.[19] According to these advocates some 5–10 per cent of Americans (that is, African Americans) would 'benefit' from such psychosurgical procedures. Yet the racist infamy of Tuskegee and even this latest gross promise of psychosurgery as a means of pacifying the ghettos were met with silence from the custodians of the ethics of biomedical research and patient care, the American Medical Association.

The memory of the outrage of Tuskegee has not retreated. African Americans remain cautious of the Public Health Service and of white American biomedical researchers.[20] It took until 2010 for another ethical horror story of the same scale as Tuskegee to be revealed. Wellesley Professor of women's studies, Susan Reverby, uncovered documents showing that in 1946 the US Public Health Service, in collaboration with Guatemalan doctor Juan Funes, who had been trained by the Service, had deliberately exposed several hundred

Guatemalan prisoners to syphilis by bringing infected prostitutes into the prisons or applying infected materials to skin and penile abrasions so as to study the course of the disease.[21] Four hundred and twenty-seven men became infected, of whom 369 were later treated with penicillin. In 1997, more than six decades after the start of the Tuskegee studies, President Clinton still found it necessary to make a public apology to the few remaining survivors, and indeed to the entire African American community. Thirteen years later, President Obama made a similar apology to the President of Guatemala.

RELIGION AND THE BIRTH OF US BIOETHICS

While the AMA seemed imperturbable, religious ethicists from the churches – both Catholic and Protestant – were not. Aware of the new ethical challenges raised by the fast-moving frontiers of biomedical research, and with their consciences stirred by the civil rights movement, they had begun to explore the relationship between the ethical control of research and public policy. The Jesuit priest Albert Jonsen, author of a history of bioethics, was one such pioneer, and the Episcopal bishop and teacher of divinity at Harvard Joseph Fletcher, another. As sociologists Renée Fox and Judith Swazey point out in their history of US bioethics, even those early bioethicists who were not clerics themselves frequently came from religious families.[22]

The establishment of the first two well-funded bioethical study centres in the US marks the point at which bioethics began to be taken seriously. The first was the Hastings Center for Bioethics and Public Policy, established in 1969, with the Catholic Daniel Callahan as its first Director; the second, two years later, was the Kennedy Institute. The Kennedy family's Catholicism guaranteed that the new reproductive technologies would be included among its wide range of ethical concerns. Senator Edward Kennedy – with his interest in biomedical research and, according to his obituarists, a need to serve lifelong penance for the murky Chappaquiddick incident in 1969 in which a young woman in a car he was driving was drowned – became its President.

Catholics were among the first to see the development of the new reproductive technologies as being likely to bring about conceptions outside the womb. If successful this would enable scientists to intervene in the reproductive process so that children could be born to hitherto infertile couples, with wide implications. Thus the Kennedy Institute's inaugural conference included sessions on 'Who shall be born?', 'Is procreation a right?' and 'Fabricated babies: the ethics of the new technologies in beginning life'. Kennedy's presence ensured a guest list for the inauguration which reads like a celebrity endorsement, running the gamut from Mother Teresa to Germaine Greer via Ivan Illich, Jim Watson, IVF pioneer Patrick Steptoe, Nobel Peace Prize winner and holocaust survivor Elie Weisel, and José Delgado, neuropsychologist and proponent of a society 'psychocivilised' by radio-controlled brain manipulation. But perhaps it was the gifts given to these honoured participants that said it all: a crystal sculpture of an angel holding a baby in her arms – and a cheque. This fusion of God and Mammon at the christening of the Kennedy Institute aptly reflected the birth of bioethics as being as much a business enterprise as an ethical code for biomedical research.

It was Kennedy too who responded to the exposure of racist abuse in the Tuskegee research on the bodies of poor black men by steering through Congress the National Research Act, establishing a Commission with a brief to determine basic bioethical principles and develop research guidelines. With the customary US distaste for regulation, Congress had turned to ethics as a means of reconfiguring research practice. The Commission included philosophers drawn from the then dominant school of analytic philosophy, as well as religious ethicists. Its Belmont Report,[23] published in 1979, became the foundational text of principlist bioethics, given wider circulation through the textbook *Principles of Biomedical Ethics*, written by Belmont's main author, the moral philosopher Tom Beauchamp, together with the religious ethicist James Childress.[24] The central assumption of Belmont is that of the autonomy of the individual – a concept integral to the individualism of American culture. On this basis the report set out two further core ethical principles for biomedical research: beneficence and justice. In their book, Beauchamp and Childress made two further moves, naming non-maleficence as a fourth principle, and expanding Belmont's agenda from research ethics to include those for patient care – that is, medical ethics.

Looking back on Belmont from the vantage point of the 1990s, bioethicist and lawyer George Annas criticised the report as defending the status quo, in that it assumed optimistically that research was inherently good, that experimentation is rarely harmful, and that researcher-dominated ethical review boards can adequately protect research subjects (all assumptions not confined to Belmont).[25] Notwithstanding the salience of Annas's criticism, principlism won the day. Despite the torrent of biomedical research in the years since their book was published, the piled-up copies of *Principles of Biomedical Ethics* to be found in every university bookshop even today suggest that principlism is not yet dead. The weakness of the presumption of a universalist ethics and the unlikelihood of their consistent, let alone just, application was to be exposed later.

THE BIOETHICS ENTERPRISE

It is the conflation of morals and money symbolised by the Kennedy Institute inauguration that marks out bioethics as a uniquely US creation. By the first decade of the twenty-first century it had become a well-financed academic practice, with clearly mapped out career paths for its practitioners, turf wars between rival bioethics centres, consultancies, and a clutch of competing print and online journals. Over the course of the four decades since its institutionalisation, bioethics has become firmly entrenched within, and a taken-for-granted feature of, biomedical research. Ethics no longer have to be internalised by individual researchers, they can, as a report in *Nature* described it, 'dial E for Ethics' and consult a bioethicist colleague on campus.[26] Any major biomedical project will today have an embedded professional ethicist on the payroll, whether as a staff member or a consultant. As sociologists Raymond de Vries and Janeardean Subedi[27] and historian of science Charles Rosenberg[28] point out, bioethics can be better understood as an enterprise than a discipline.** The rise of the bioethicists can be seen as the outsourcing of ethics – a very neoliberal practice.

Professional bioethics' preoccupation with 'autonomy' fits well with the individualism which, as de Tocqueville observed in the early nineteenth century, sits at the heart of American society, alongside its energetic commitment to democracy. Even though in later versions of Beauchamp and Childress's theorising, autonomy has been modified to stress respect for autonomous individuals, those individuals cannot escape the structural constraints of their position within a harshly hierarchical society. For example, the justice principle takes into consideration only those justice issues that relate to US citizens with health insurance. As feminist philosopher Rosemary Tonge[29] pointed out in 1997, this shrivelled concept of justice has nothing to say about those millions who, despite Obama's reforms, remain outside the nation's health-care system.

Feminist philosophers have laid siege to principlism. Anne Donchin stresses that Beauchamp and Childress's autonomous individuals are in reality embedded in a mesh of interdependent social relations. The intersections of family, gender, class and race, to name just a few of the multiple social divisions, are ignored by principlism with its preoccupation with the abstraction of the individual independent of social context.[30] Sara Ruddick[31] and Carol Gilligan[32] have sought to go beyond both consequentialist ethics and deontological or rule-based ethics with their emphasis on dependency and interdependency, underpinning their ethic of care. Mainstream bioethics has accommodated this feminist preoccupation with interdependency, though rarely crediting its authors.

Just as Beauchamp and Childress's abstract concept of autonomy ran into difficulties in research, so too has the Helsinki concept of informed consent, originally intended to emphasise the importance of transmitting reliable and accurate information to the potential research subject. This was never easy, and with the huge growth of

** The proceedings of a European meeting on bioethics Steven attended were disrupted by an extended squabble between the members of two US rival bioethics institutes over the 'ownership' of neuroethics, a rapidly expanding subfield of bioethics.

genomics over the last half-century, has become difficult or impossible. If a person consents to genetic testing, the findings may relate not only to that individual but also to their genetic kin – parents, siblings, children – who may not want the information, or want it known. Twenty-first century genomics subscribes to the claim of the sixteenth-century poet John Donne that 'No man is an island'. Furthermore, if DNA samples are stored – for instance in one of the rapidly expanding biobanks – they may subsequently be used for tests and procedures not covered in the original consent form. In these situations there are, to quote the unlikely figure of Donald Rumsfeld, both known unknowns and unknown unknowns. Being able to provide adequate information concerning possible risk, which is essential to such a cognitively driven concept as informed consent, here verges on the impossible.

With the geneticists finding that the individualist notion of informed consent was increasingly unusable in their research, above all as the huge DNA biobanks began to emerge at the millennium, the ethicists have turned to collective forms of consent for a way through the impasse. Far from new to anthropologists and central in the conflicts over the Human Genome Diversity Project, collective consent is part and parcel of securing access for any researcher seeking to study indigenous peoples – but just who forms the collective and who that collective represents in actuality can be opaque. One early example of the collectivist turn was proposed by the English philosopher Ruth Chadwick and the Canadian lawyer Bertha Knoppers. They suggest five new principles: reciprocity, mutuality, solidarity, citizenry and universality[33] – this last intended to take account of cultural differences and transcend the Westernocentric view of principlist bioethics. But all their principles set aside the relations of power. Just how does a poor Aboriginal community establish reciprocity with the British government. Who is to determine what is mutual – the powerful or the weak? Does their principle of solidarity, defined as solidarity for a common good, tell us who is to define that common good? It might require, as with Marx in the nineteenth century, or the Occupy movement of today, an overturning of the existing social order. The concept of the citizen is in difficulty in the context of the immense population movements that are a consequence of globalisation. The patient building from below of multiculturalism, embracing citizens and non-citizens (immigrants and refugees) alike, represents a new collectivity that transcends citizenship. The bioethicists' hope that universality can take account of cultural differences sets aside both the current reality that the driving force of biotechnology comes from EuroAmerica and that such Westernocentrism may well be superseded in the coming decades by the new biotech giants of the East.

PROPERTY IN THE BODY

Until relatively recently there was an unquestioning assumption that surgical waste, including post-mortem body parts, became the property of the surgeon or the hospital. Surgeons could and did amass substantial private collections, often for research and teaching but sometimes out of little more than curiosity. The skeleton of the 'Irish Giant' Charles Byrne, 7' 7" (230 cm) tall, who was so horrified at the thought of being dissected as a freak after his death that he expressed his wish to be buried at sea, was purchased by the leading surgeon John Hunter in 1783 for the then huge sum of £500. It was placed in the Hunterian Museum of the Royal College of Surgeons in London. Today, recurated with great elegance, the once exclusive pathology museum is open to the public. Byrne's skeleton is still on view, technically legal but still morally deplorable, even though many of the human skulls that were originally part of the collection have been relocated out of public view to the basements of London's Natural History Museum.

Widespread resistance to this old concept of clinical property in the body came dramatically into public view in Britain during the 1990s with a series of scandals concerning hospitals', and even individual doctors', retention of body parts. Following post-mortems of children who had died from cardiac arrest in the hospital, the Bristol Royal Infirmary had removed their hearts for teaching and research. For medical culture, these were 'waste', and their disposal raised no ethical issues. This was not how it was seen by the parents, who insisted that they should have been consulted. The resulting moral scandal was massively amplified by the media, and was soon followed by that of Alder Hey Hospital in Liverpool. In January 2001 the official Alder Hey report confirmed the media stories and spelt out in detail just what had taken place. The pathologist Dick van Velzen had systematically ordered the 'unethical ... stripping of every organ from every child who had had a postmortem' during his time at the hospital, and indeed took them with him when he moved to another hospital. He had ordered these even for those children whose parents had specifically stated that they did not want a full post-mortem. Looking at this practice across the hospital system, the report revealed that over 104,000 organs, body parts and entire bodies of foetuses and still-born babies were stored in 210 NHS facilities. Additionally 480,600 samples of tissue taken from dead patients were also being held. Furthermore, it emerged that both Birmingham Children's Hospital and Alder Hey Children's Hospital had given thymus glands removed from live children during heart surgery to a pharmaceutical company for research, in return for financial 'donations'.

Initially the government supported traditional medical culture, but the public and many doctors resisted, feeling that it was time to change. Hospital culture and practice claimed, that the institution was the moral custodian of 'waste' body parts of potential use for teaching or research, not the individual clinician – so van Velzen had breached medical culture by assuming that the human tissue was 'his' and that therefore he had the right to take the collection with him when moving to another hospital. But the government position could not hold in the face of rising public anger and a committee of enquiry was established leading to the Human Tissue Act in 2004. This regulates the removal, storage and use of human tissue and stipulates the need for informed consent except in the case of DNA, that crucial twenty-first-century commodity. With these new oppositional sensibilities, property in the body was de facto returned to the parents of those babies whose organs had been taken. But although this was a success, the struggle over property in the body continues.

While there is no evidence that van Velzen intended to profit from his sequestration of the body parts, two American cases have brought the question of ownership sharply into view, and have been repeatedly returned to by those concerned with the ethics of research. The first dates from 1951, when Henrietta Lacks, a poor African American woman suffering from cervical cancer, entered the Johns Hopkins Hospital in Baltimore. During her treatment with radium, some of the cancerous cells from her cervix were removed by her clinician without her consent. They were cultured by the tissue culture specialist George Gey and were found to have very unusual properties. Unlike previous cultured cells, Henrietta's continued to grow and divide indefinitely, providing a major research resource. Under the name HeLa, the cell lines, initially distributed by Gey free and unpatented to other researchers, were commercialised. Henrietta Lacks died the same year, but her impoverished family has to this day received neither recognition nor financial compensation.

The second case was that of John Moore, who in 1976 underwent treatment for leukaemia under the care of David Golde at the University of California Medical Center. As in the Lacks case, Golde derived a cell line from Moore's excised tissue, which the University of California patented under the name Mo cell line. Entering into an agreement with the Genetics Institute, a Harvard-based biotech company, Golde became a paid consultant and acquired the rights to 75,000 shares of common stock. The Genetics Institute also agreed to pay Golde and the university at least $330,000 over three years, including a pro-rata share of Golde's salary and fringe benefits, in exchange for exclusive access to the materials and research performed on the cell line and products derived from it. On 4 June 1982, the drug company Sandoz was added to the agreement, and compensation payable to Golde and the Regents was increased by $110,000. When Moore discovered this, he sued Golde. However, the California Supreme Court denied him a property right in his cells, in part on the grounds that the cell line and the products derived from it were factually and legally distinct from his body. However, the court concluded that Golde did have an obligation to reveal his financial interest in the materials harvested from Moore, and that the latter would be allowed to bring a claim for any injury that he sustained as a result of the physician's failure to disclose those circumstances. In reaction to this judgment, some patient groups, such as those in the US with Canavan disease (an early onset genetic disorder primarily affecting the brain), have banded together to negotiate their consent for tissue removal against a share in patent and intellectual property rights and hence of any profit. Within neoliberalism trading property in the body, including the diseased body, makes sense.[34]

TRIALLING GENES AND DRUGS

The scandals surrounding gene therapy trials have generated the sharpest debate. The extreme hype of the molecular biologists over the prospect of gene therapy has played no small part in focusing media and public attention on its successes and failures. Less attention was initially placed by the media on the problem of the ethical control of research on human subjects. The example of just two trials, one American and one French, serve to demonstrate the difference between effective and ineffective controls. The first is the highly publicised US case of the research-inflicted death of Jesse Gelsinger in 1999. Gelsinger, an eighteen-year-old with a serious genetic condition resulting in liver disease, was recruited into a gene therapy trial run by the University of Pennsylvania. To the horror of his family – given that he was reasonably well when he entered the trial – the gene therapy killed him. The presence of the eminent academic bioethicist Arthur Caplan on the university team, together with the usual procedure of informed consent, had given insufficient protection. The subsequent formal investigation concluded that the university was at fault: first, inclusion of Gelsinger despite it not being clinically appropriate; second, failure to report that two patients had earlier experienced serious side effects from the gene therapy; and third, failure to mention, as part of the informed consent process, the deaths of monkeys given a similar treatment. The lead researcher lost his job and a much tighter framework for future gene therapy trials was put in place. Gelsinger's family sued but the case was settled out of court. By contrast, in the French trial of gene therapy for SCID (severe combined immunodeficiency, or 'bubble babies'), in 2002, when two of the eleven infants in the trial died from leukaemia, the treatment was stopped by the research team and the trial terminated. The better ethics of the French team were evident to all. Nonetheless, the failure of gene therapy trials, however ethical, meant that they were abandoned for the rest of the decade. Then they were cautiously resumed, with good results reported by 2011 for forty-three out of fifty-seven children in two SCID trials and twenty-eight out of thirty for a genetic form of blindness.[35]

For new drugs to be licensed by regulatory bodies such as the Food and Drug Administration in the US and the Commission on Human Medicines in the UK they have to pass through a series of trials, first on animals, then on healthy human subjects, followed by a pilot and finally a large-scale trial on patients. While at every stage the quality of the research must be scrutinised, it is the final trial that poses the greatest risk to patient subjects and is the point where most drugs fail. The failure of the European regulatory bodies (though not of those in the US) to pick up on the poor quality of the research underpinning thalidomide points to the critical importance of the work of such bodies. The clinical trial phase requires many thousands of treatment-naive subjects, that is, people not taking any relevant medication. In the US the lack of a universal health-care system has created a market opportunity for the pharmaceutical companies. Money is a sufficient attraction for many poor but healthy people, but for others who may be facing near certain death, taking part in a trial may be the only means of obtaining access to potentially life-saving drugs. Under these circumstances informed consent may mean little more than signing on the dotted line.[36] For some healthy volunteers the pay is inducement enough to compensate for the risks involved, although such trials can go horribly wrong, as in the case of the young volunteers at Britain's Northwick

Park Hospital in 2006 who suffered nearly fatal immunological collapse after the injection of an anti-inflammatory drug intended for leukaemia treatment.[37]

Finding sufficient numbers for clinical trials has become progressively harder and more costly in the richer countries, especially since for many trials the subjects are required to be 'drug-naive'. The cost per subject in the US is now about $20,000 – carried out in what is described as a 'second class' centre; the cost in a 'first class' centre in India, by contrast, is around $1,000–1,500 per subject. These financial pressures have led to the global outsourcing of clinical trials – ostensibly carried out to the same technical level and to the same ethical standards. However, the US Food and Drug Administration checks that these are in place for less than 1 per cent of foreign trials, where well-founded doubts can lie. The long history of testing contraceptives on women in the poorer countries of the South sets the precedent for Big Pharma. Nor is this a one-way process; post-communist governments, such as that of Estonia, while not exactly selling their populations, do advertise them as ideal research subjects.[38] This outsourcing is carried out by the new contract research organisations, and is now very big business – currently no fewer than 78 per cent of all US and European patient trials are conducted abroad, mostly in poorer countries. Of the many thousands of clinical trials currently under way, the largest number are in China, with India and Russia not far behind.

In the process, Helsinki and Nuremberg have been set aside. In 2000 the *Washington Post* broke the story of a 1996 experiment conducted by Pfizer researchers in Kano, Nigeria, during a major meningitis epidemic. The trial resulted in the death, among others, of a ten-year-old girl known only as Subject 6587–0069. The researchers had monitored her dying without modifying the treatment. When the families brought a court case against Pfizer, the company argued, initially successfully, that there were no international norms requiring informed consent. Such trials, the *Post* reported, were 'poorly regulated' and 'dominated by private interests'. It took a re-trial in 2009 for a US court to reaffirm the Nuremberg Code as foundational and of universal application.[39]

MATER SEMPER CERTA EST?

The moral crisis posed by the birth of Louise Brown, the world's first in vitro fertilisation (IVF) baby, in Oldham in 1978, dramatically underscored the need not just for bioethics but for legally enforceable regulation of biomedical research. The baby's birth was the realisation of speculations by scientists and science-fiction writers, from J.D. Bernal to Aldous Huxley, since at least the 1930s. As foreseen by Huxley in the opening pages of *Brave New World*, Louise was the result of her parents' gametes being fused in a Petri dish. Hailed as a technological marvel and a solution to problems of infertility, her birth, almost inadvertently, erased the *mater semper certa est* of Roman law.

IVF added to the old uncertainties about the identity of the biological father that of the identity of the biological mother. The possibility of one woman providing the gamete and another the womb challenged everyday understanding. Now the 'mother' could be two or even three different women: the provider of the egg, the carrier of the foetus, and the carer of the child. Thus although the identity of the woman who carries the pregnancy remains certain, together with that of the carer, the identity of the woman who provided the egg will have been anonymised. A complexity of identity which only increases as human embryology advances.

The announcement of Louise Brown's birth startled the world, even though, as the inaugural Kennedy Institute meeting had recognised eight years previously, the realistic possibility of a 'test tube baby' was central to the ethical and social agenda – especially to one funded by Catholics. This Catholic preoccupation was shared by those biologists aware of the developments in human embryology. In the same year as the Kennedy Institute meeting, a conference on the social impact of modern biology was organised in London by a group of life scientists chaired by Nobel Laureate and President of the British Society for Social Responsibility in Science, Maurice Wilkins.[40] Among the speakers was Cambridge embryologist Robert Edwards, whose collaboration with the gynaecologist Patrick Steptoe on IVF looked to be nearing realisation. Edwards's presentation recognised the cultural and ethical sensitivity about research on human fertility, and set out both in his talk and in later interventions his proposals for ensuring that IVF was well regulated and hence socially acceptable. But beneath the desire to help infertile couples lay his eugenic goals, and equally clearly his judgement on who was to take the decisions about which couples to help: 'to improve the quality of human life by regulation of fertility and by identifying and averting the causes of certain forms of anomalous development, we hope to make it possible for some infertile parents to have their own children wherever we believe our help to be valid and meaningful'. That biomedical researchers should be the gatekeepers was entirely self-evident. However, only when Louise Brown was born did Edwards and Steptoe's achievement become a global news story. Headlines asked: 'Are scientists playing God?' 'Is this the end of natural conception?' 'Will babies be cloned?', the speculations amplifying a moral crisis which demanded political as well as media attention.

It took the government four years from Louise's birth to appoint moral philosopher Dame Mary Warnock to chair a committee of enquiry with the brief to consider the implications of developments in IVF and embryological research and to advise the government on the ethical and regulatory management of both.[41] Within Britain, committees dealing with issues touching personal and moral matters normally include representatives of faith groups and, on any science-based issue, expert natural scientists. The Warnock Committee was no exception, but for the first – and so far only – time in the history of British Royal Commissions the membership included near equal numbers of women and men. This was all the more astonishing in that it had been set up by a Thatcher government hostile to feminism. Warnock herself was a conservative who had previously chaired a committee of enquiry and could therefore be regarded as a safe pair of hands. However, despite the silent nod to an increasingly strong women's movement no feminists as such were appointed. The Committee member most sympathetic to feminism was biologist Anne McLaren. The converse of Warnock, McLaren was left-wing, and already an FRS for her work in embryology.

While the Committee was still deliberating, McLaren took part in feminist meetings discussing the significance for women of the new reproductive technologies. For feminists the birth of Louise raised the question of who gets to decide who will be mother, and, given the rapid growth of molecular genetics and intensified antenatal testing, who decides which foetuses will survive. Seventies feminists feared that the medical patriarchs would, left unopposed, be all too happy to step forward as the decision makers. The strongest opposition to the new technologies came from the Feminist International Network of Resistance to Reproductive and Genetic Engineering, but even this network of radical feminists, for all its energy, did not get going until six years after Louise's birth. By then the horse was well and truly out of the stable. By 2011, three million babies had been born through IVF worldwide.

The final report, signed by the majority of the Warnock Committee, welcomed, with caveats, the development of IVF and of experimentation on human embryos up to a limit of fourteen days from fertilisation. Acknowledging the need to allay public anxiety, the committee recommended that the government establish a Human Fertilisation and Embryology Authority (HFEA) with appropriate regulatory powers. The fourteen-day rule more than satisfied the embryologists, as there were technical problems in keeping the developing embryo alive in vitro for even that length of time. This time limit could be given some ethical legitimacy by the argument that it was at this point in embryonic development that a precursor to the nervous system – the neural streak – could be observed. The fourteen-day milestone came to be a secular version of the centuries-long Christian debate over the moment at which the soul entered the foetus. Thus despite science's traditional claim that scientific knowledge has nothing to say about moral values, it was developmental biology (almost certainly in the person of Anne McLaren) that wrote the moral economy of the embryo. The move was made with such skill that the advent of the neural streak as marking the first moment of potential personhood was received as entirely 'natural'.

The case is a near classic example of science as culture. The moral economy, 'written by nature' as interpreted by biologists, was so successful that most European countries with equivalent research programmes adopted the fourteen-day limit within their own regulatory frameworks. The political success of Warnock's committee was less than celebrated by other philosophers, with Rom Harré complaining that a golden opportunity had been missed to put bioethics on a philosophically sound basis. The complaint was not without foundation; philosophers had played an important part in the US Belmont Report, and in Britain Harré and Elizabeth Anscombe had made important contributions to moral philosophy. Why was philosophical reflection excluded? British political culture was pragmatic; Warnock understood this and produced an entirely workable compromise. The life scientists were content: research could proceed. Germany, then implacably opposed to human embryological research, was the conspicuous exception. By contrast, despite the early discussions at the Kennedy Institute, the US introduced little or no regulation, its political powers divided between the Federal government and the States. Not until the religious fundamentalist George W. Bush took office was Federal funding denied to research involving human embryos.

NEW ETHICAL COMMITTEES IN THE GLOBAL ECONOMY

By the millennium, the governments of most countries with medical research programmes had established national bioethics committees, linked to European and UNESCO committees. The oddity is that Britain has not. The Nuffield Council on Bioethics, set up in 1991, is neither a governmental body nor attached to a national academy, but is instead a private institution, originally funded by the Nuffield Foundation, but now largely by the Wellcome Trust and the Medical Research Council. It has no democratic accountability and its governing Council appoints ad hoc committees to examine biomedical advances thought to raise moral and social issues. The Council says that it has no commitment to any one philosophical approach to bioethics, specifically not to principlism or human dignity theory. Despite this eclecticism, their reports suggest a feel for what is socially/politically acceptable, and the ghostly presence of Jeremy Bentham.

The French took a more intellectually courageous approach when President Mitterrand established the world's first permanent National Ethical Commission in 1983. Recognising the diversity of philosophical and religious traditions, its initial membership of thirty-seven included members from five principal faiths and philosophies: Catholicism, Protestantism, Judaism, Islam – and Marxism. This acknowledgement of the strong French cultural tradition of a materialist and secular viewpoint on ethical matters is distinctive. The Swedish National Council on Medical Ethics includes among its members representatives of all the parties in Parliament, where women are strongly represented, along with philosophers, lawyers, doctors, nurses and activists from the disability movement.

Germany was slower to establish a national council, despite the vigorous bioethical debates over the decades among the medical profession and at Land level. In 1984 the ministries of research and legal affairs set up ad hoc committees to make recommendations for legislation on IVF and gene therapy, followed by a parliamentary commission on genetic engineering, and, in 1988, one on reproductive medicine. In 1991, a law was passed forbidding any form of human embryo research. But it wasn't until 2001 that a National Ethics Council was established, with twenty-five members – 'prominent representatives of the scientific, medical, theological, philosophical, social, legal, ecological and economic worlds' nominated by the Chancellor – including activist intellectuals such as the molecular biologist and prominent Green feminist Regine Kollek. Germany's historic sensitivities to eugenics and medical experimentation, together with its strong philosophical tradition of human dignity ethics, have led the Council to produce some of the most critical bioethical opinions. Other European countries have established variants of these several structures, as has the Council of Europe itself, which in 1997 formulated a code based on Helsinki but with legal powers, namely the Oviedo Convention on the Protection of Human Rights and Dignity in Biomedicine, which includes a prohibition on human cloning and restrictions on transplantation.

THE HUMAN GENOME PROJECT AND ELSI

As mentioned in Chapter 1, the huge expansion of US bioethics can be dated to the launch of the HGP in 1988, when Jim Watson, its first director, announced out of the blue that there was to be a richly funded programme studying its ethical legal and social implications (ELSI). Watson's invention of ELSI showed that he was shrewd enough to recognise that the HGP was likely to come under attack for reviving the dormant spectre of eugenics if it made no serious attempt to acknowledge the dangers involved. Neuropsychologist Nancy Wexler, herself at risk from Huntington's disease (HD), a single-gene degenerative neurological disorder, was appointed ELSI's first Chair. She had been part of a research team going to a region of Venezuela with a high incidence of HD to recruit research subjects from at risk families. (As we discuss in Chapter 4, although the HD gene was identified in 1983 there is still, in 2012, no therapy). Appointing Nancy Wexler was thus more than just a smart move. She was a scientist people could trust, combining a personal understanding of those genetically at risk with direct participation in the gene-hunting which she and her colleagues believed would lead to treatment.

ELSI ran from the inception of the HGP until its second 'completion' in 2003 and was the world's largest bioethics response to the challenge of the social and legal implications of molecular genetics, disbursing funds to ethicists, lawyers and social scientists across many US academic institutions. The acronym has persisted well beyond the lifetime of the original project, although Europeans have changed the I to an A for Aspects to suggest a less technological determinist approach. It was not until 1995 however, several years after most European countries, that President Clinton established a National Bioethics Advisory Commission (NBAC) with the brief of proposing regulations to prevent the abuse of human subjects in federally funded research. Harold Shapiro, a Canadian economist and President of Princeton University, was appointed as Chair (his twin brother, at McGill in Montreal, took on the matching task for the Canadians).

The Clinton Commission had barely been established when, in 1996, along came the birth of Dolly the sheep – the first cloned mammal. If the birth of Louise provoked moral anxiety, Dolly's birth precipitated a near global moral panic. Had we suddenly found ourselves inside the scientific dystopia of Huxley's brave new world? Was biotechnology soon to make possible the nightmare scenario of humans cloned in the laboratory, constructed by biotechnologists? (The nearest comparable research project, which fortunately never proceeded beyond the proposal stage, was made in the 1930s by Soviet scientists who proposed to mate male great apes with African women – to advance evolutionary theory.) Ian Wilmutt's team, based at the Agricultural Research Council's laboratories in Roslin in Scotland, had brought off an extraordinary scientific achievement which, unlike the birth of Louise Brown, not even the biological community had anticipated.[42] Clinton responded before the Commission had reported, signing a temporary order forbidding federal funds for human cloning research. He then turned to the NBAC, ordering them to report within ninety days on what should be done. NBAC members had conflicting views on the religious, scientific and ethical issues, but on safety grounds took the unanimous view that there should be a legislative ban on human cloning, with a five-year sunset clause.

In 2001 the incoming Republican President, George W. Bush, abolished the Clinton Commission and established a new body, the President's Council on Bioethics. The new Council in its composition and choice of study topics reflected Bush's concerns over embryo and stem cell research, assisted reproduction, cloning, neurological 'enhancement' and 'end-of-life' issues. Its eighteen members were formally described as drawn from science, medicine, philosophy, theology and social sciences. This did not, however, prevent vocal criticism from the US science community. *Science* complained that the Council was 'stacked' – politically biased in its opposition to, among other matters, stem cell research based on the use of human embryos. Its Chair from 2001 to 2005, the Chicago bioethicist Leo Kass, was on record as an opponent of human cloning and euthanasia, and critical of unrestrained technological progress and embryo research. He was succeeded by Edmund Pellegrino, past President of the Catholic University of America.

The invariably political nature of such appointments was once again demonstrated by the immediate changes introduced by Barack Obama. He replaced Bush's Council with a Presidential Commission for the Study of Bioethical Issues, with a new less religious and more pro-science membership. The Commission was given a new brief to examine 'the intersection of science with human rights' and the ethical legal and social issues arising from developments in biomedical research. Indicative of the rapid advances in research in the decade since the previous Commissions had been established, there was now to be a focus on the neurosciences, synthetic biology and robotics. Like Clinton before him, Obama did not wait for the new Commission to report before he revoked Bush's ban on federally funded stem cell research.

GLOBALISING BIOETHICS

That generous project for bioethics that originated in Weimar Germany was concerned with the well-being of both nature and society in an increasingly scientific and technological age, and remains by far the most politically attractive and intellectually coherent proposal of its kind. Jahr's invocation of a future in which children would no longer pick flowers only to throw them away envisions a moral growth in humanity's respect for nature. For a time the double agenda of bioethics and biopolitics was sustained by the Green movement and in particular by the German Greens. They took on the issues of new reproductive technologies and genetically modified organisms with tremendous energy, bringing together in their heyday near unimaginable alliances between ecologists, feminists, Christians and Marxists. But that heyday has passed and the political defence of humans has weakened – today the Green movement primarily focuses on nature. Meanwhile the bioethics of biomedicine in this now massively scientific and technological age has, despite religious (pace Bush) and feminist opposition, become increasingly professionalised and outsourced.

Scandals, ethics, religion, culture and politics have shaped the professionalised discourse of bioethics, but, just as it did for Darwin and the physiologists, the drive to translate the brute facts of life itself into the language of science remains central to late modernity. For organised knowledge to proceed, both science and the state have to accommodate moderate ethical opposition while fending off more radical opposition – whether to human genomics as initially in the case of Germany and the European Parliament, or to animal research as in case of the animal rights movement. In the past, each nation state established its own model of regulation, but with the globalisation of technoscience what was national, and for Europe continental, is now a global matter. As long as EuroAmerica dominates biomedical research, for the global flow of products including intellectual property to move smoothly, the emerging research giants of East Asia must adapt their bioethics to those of these old research giants. This convergence of national bioethical councils and policies is both a reflection of, and a response to, the globalisation of biomedical research. Viewed positively it is an indication of the global recognition of the need for bioethics; viewed rather more critically it is all about the facilitation of research through the standardisation of social acceptability, thus ensuring the smooth marketability of biomedical commodities in a global economy still dominated by the West. What happens to bioethics when the technosciences' centre of production leaves EuroAmerica is an open question.

4

From State to Consumer Eugenics

THE CENTURY OF EUGENICS

Thinking about human genetics today, not least about the vaunted possibility of the 'perfect baby', we seem to be both near to and very distant from the spectre of eugenics which haunted the twentieth century, above all in the hideous episode of Nazism. This new science fiction in which prospective parents get to choose the eye colour, height, intelligence, looks and disease resistance of their proposed baby is part of a new consumer culture without limits. If you want it, can pay for it, and someone can provide it, then it, whatever 'it' is, is yours. A revitalised economic liberalism enthrones the consumer as king – or even queen. Of course, accompanying these new technological possibilities there will be some moral questioning about the desirability of introducing the market into parenthood, but with a legal and ethical framework weakened by our capitulation to the market as the chief arbiter of our futures, this is unlikely to have much force. Meanwhile, even though tasteless, absurd, and even impossible, the dream of the perfect baby made workable by the promethean promises of the biotechnosciences takes its place alongside other consumer fantasies of the perfect partner, house, job, garden or suit. Perhaps such unrealisable fantasies are connected to what the World Health Organisation describes as the current worldwide pandemic levels of depression.

The Century of the Gene was the title Evelyn Fox Keller gave to her account of the role of genetics in building the organism over the course of the twentieth century.[1] Nonetheless, despite the anger of geneticists that their science could have anything to do with the pseudoscience of eugenics, that same twentieth century could be equally named 'the century of eugenics'; a point Keller glosses over in her somewhat Whiggish account. Eugenics and genetics have had, like conjoint twins, both individual and linked histories. Genetics as the science of biological difference at last creates the possibility of realising Francis Galton's nineteenth-century biopolitical project of eugenics. With the recovery of Mendel's work at the beginning of the twentieth century, the concept of the gene and the birth of the discipline of genetics, which offered a mechanism for the transmission of difference, could now be recruited to eugenics. With the new authority of genetics, biomedical science increasingly saw itself as able to distinguish scientifically the innately wellborn from the innately unfit. For the sake of the fitness of the nation the birth of the former was to be encouraged (positive eugenics) and the latter discouraged (negative eugenics). Darwinian evolution, with its central notions of fitness and natural and artificial selection, was crucial in building support for the eugenic project, as it played into how nations conceived themselves, particularly as 'races'. Nations as races stood in competition with one another and thus needed to ensure the best quality human stock and to minimise the numbers of the unfit. Scientific eugenics could achieve this as surely as Darwin's pigeon breeders could improve their birds by what he called artificial selection.

Mendelism provided the early eugenicists with a simple model: criminality, feeble-mindedness and moral turpitude were seen as single heritable characters which could be eliminated from the population by preventing those with such traits from reproducing. As judge Oliver Wendell Holmes famously remarked in 1927, in the landmark case of Carrie Buck, a poor white woman, which legitimised compulsory sterilisation in the US: 'We have seen more than once that the public welfare may call upon the best citizens for their lives. It would be strange if it could not call upon those who already sap the strength of the State for these lesser sacrifices, often not felt to be such by those concerned, to prevent our being swamped with incompetence,' concluding, 'three generations of imbeciles is enough'.[2]

The very title of US geneticist William Castle's textbook *Genetics and Eugenics*, first published in 1916, is a classic statement of the conjoint twins. It ran to four editions over the next fifteen years and became the standard text in both Ivy League and the state universities, hence a major influence on several generations of biomedical students and significant in giving scientific legitimation to the eugenic movement. Another key source of scientific legitimacy was provided by the powerhouse of US eugenics, the Eugenic Record Office at Cold Spring Harbor, directed by Charles Davenport and Harry Laughlin and funded generously by the Carnegie Foundation. There the researchers and their volunteer helpers collected tens of thousands of genealogies from across the US to identify the heritability of traits, from temperament to polydactyly (multiple fingers). Given the federal structure of the US, the extent of eugenic practice was both more variable and harder to document than in more centralised and smaller countries. During the 1930s over thirty States passed eugenic sterilisation laws for those judged to be anything from moronic or sexually deviant to alcoholic or even deaf or blind. By 1935 more than 45,000 eugenic sterilisations had been carried out – more than half of them in California. The unregulated powers of the doctors in charge of the institutions for the unfit gave them considerable latitude as to whether they would or would not sterilise those in their care. As with the Nazi doctors, the science of eugenics endowed their actions with a sense of civic duty and best biomedical practice. The combination of the legacy of slavery, eugenics and race theory ensured that African Americans, and particularly women, were disproportionately classified as unfit and disproportionately sterilised. As late as the 1970s young black women who had become mothers outside of marriage were still being forcibly sterilised for being self-evidently morally unfit. It took the rise of the civil rights and women's movements to expose the programme and eventually force its termination.

Although geneticists were strongly represented among the eugenicists, there were considerable differences between them from the very beginning of the discipline. Many, including Wilhelm Johannsen, one of the three re-discoverers of Mendel's work at the start of the twentieth century, insisted that things were more complex. Recessive characters, present in the genotype, might not always be expressed in the phenotype. Furthermore, to eliminate such characters from the population by sterilising their carriers, even if these could be identified and the character were the result of a single mutation, would take many generations – even as many as a thousand. William Bateson shared Johannsen's recognition of the complexity. However, contradictorily, he feared that it might be employed in less moral hands than those of the British. Bateson's nationalist moral superiority was a foretaste of that of Ivy and Alexander who, having played a key part in formulating the Nuremberg Code, with equal nationalism saw US biomedical researchers as being automatically ethical. The US experimental geneticist and pragmatist philosopher Herbert Jennings was less complacent; horrified by the fast-growing eugenic movement of the 1920s with its racist commitment to nativism, he gave up his research to prioritise campaigning against it.

The Depression and the suffering of the unemployed saw the scientists divided between left and right (with the adoption of Marxism by many of the most brilliant Cambridge students, including the natural scientists). When in the 1930s the mathematical geneticists replaced simplistic Mendelism by integrating it with Darwinian natural selection to create the Modern Synthesis, the already existing political divisions among the geneticists intensified. Some, like the deeply conservative Ronald Fisher and Karl Pearson (who thought Hitler had some sensible ideas about eugenics), remained faithful to what Daniel Kevles in his history of Anglo-American eugenics calls 'mainstream' eugenics – including both positive and negative eugenics. Marxists like J.B.S. Haldane and the mathematician Lancelot Hogben, and the liberal Julian Huxley, recognising that phenotypic characters were often polygenic or even non-genetic, although still committed to the eugenic project as improving human stock, saw it being achieved through improving peoples' living conditions, which would lead to the voluntary reduction of family size. But such niceties among the geneticists were largely missed by the wider intellectual and political classes.

Social amnesia on the part of twenty-first-century intellectuals prevents them recalling this huge enthusiasm for eugenics in the Protestant-dominated industrial countries, from social democratic to liberal capitalist. In the early twentieth century it would be true to say that, barring the Catholics, eugenics commanded the support of most EuroAmerican intellectuals – not just racists and reactionaries but feminists, reformers and Marxists. German doctors and biologists saw the Nazis as facilitating the scientific practice of eugenics, their election to power in 1933 making it possible to introduce eugenics into Catholic Bavaria. The doctors organised the sterilisation programme in the name of the eugenic theories provided by the biologists. Even German Jewish doctors such as the psychiatrist Franz Kallmann, an advocate of sterilising the relatives of schizophrenics, continued his eugenic enthusiasms after fleeing the Nazis in 1936 for the safety of New York. Among notable feminists, the birth-control pioneers, the American Margaret Sanger and the British Marie Stopes, were passionate eugenicists. Sanger like all too many US feminists was sympathetic to racism. Stopes, in Britain, was more class oriented. Thus while the presence of their birth-control clinics brought freedom from fear of unwanted births to many poor and working-class as well as middle-class women, the motivation of these pioneers was, as US feminist historian Linda Gordon documents, more than just feminist.[3]

The playwright and Fabian socialist George Bernard Shaw was a convinced eugenicist, as were Alva and Gunnar Myrdal, the theorists of the social democratic Swedish welfare state. For the Myrdals eugenics was integral to the always nationalist project of the welfare state – without state intervention and compulsory sterilisation, the new welfare services would be overwhelmed by the increase of the feeble-minded and degenerates in what was still a poor country. Without eugenics, the social contract of the welfare state, between state capital and labour, could not work. Their thinking, when the chips came down, was not that far from that of the much pilloried Wendell Holmes.

The historian André Pichot argues that the ideology of national improvement through state eugenics was so widely accepted that the first wave of Nazi eugenics – involving mass sterilisation followed later by extermination of the mentally impaired and sick – went on largely unremarked by the Western intelligentsia.[4] Genetics and eugenics came even closer with the publication of *Human Heredity* by the leading German geneticists and race theorists Erwin Baur, Eugen Fischer and Fritz Lenz, setting out their geneticised theory of racial psychology and their endorsement of eugenics as racial hygiene.[5] Published in 1927 and translated into English in 1931, it soon became an internationally used textbook, influencing generations of students. In Germany it was seized on by Hitler as giving the authority of science to the policy of the mass extermination of the less than human – above all, but not only, the Jews. It was only then that geneticists, and particularly the British who at the time dominated the field, began to distance themselves from scientific racism, though never entirely disowning the concept of eugenics.

For the Scandinavians and the Nazis, compulsory sterilisation became the chief means of limiting the numbers of mentally and morally unfit. Could Galton, the idealistic theorist of eugenics – that science of the wellborn – have foreseen that the practice of Scandinavian eugenics would have been so harsh that it involved 'a blurred use of authoritarian measures and anonymous violence'?[6] The support of Western intellectuals for eugenics did not weaken when the Nazis secured power. During the war years the economist John Maynard Keynes and the biologist Julian Huxley were leading figures in the British Eugenics Society. Even as knowledge of the death camps grew and the price of Nazi race science became evident to those who wished to see, the Society's membership still included William Beveridge, architect of the British welfare state, Richard Titmuss, a leading figure within social policy, psychologist and IQ theorist Cyril Burt, and immunologist and later Nobel Laureate Peter Medawar. Such eugenicist histories and enthusiasms have been, if not actually hidden, at least distinctly underplayed in the national UK self-portrayal, leaving the Nazi episode to stand out as a singular horror story rather than as the monstrous epitome of a widespread ideology and its associated practices. Consequently the critique of 'state eugenics' typically addresses the Nazi horror rather than this much more pervasive state eugenicism which, despite the efforts of the historians, remains uneasily half-silenced, half-known. As the repressed rather than the confronted, such histories have returned slowly and with shame. The British and Dutch practice of sexual segregation and incarceration – which was terminated only in the mid-1970s along with forced sterilisation in Scandinavia and the US – has still to receive full recognition as eugenic, even though this was self-evident to its advocates, including Darwin's son Leonard, the President of the Eugenic Society. He was extremely clear that this was the only appropriately English way to proceed. Galton himself had been rather more utopian, hoping that the unfit would, as a matter of conscience, decide not to have children.

The Soviet Union was the exception. The contrast between Western Marxist support for eugenics and the hostility of the Soviet Union was dramatic. While Stalin had no qualms about massacring hundreds of thousands of kulaks, among other Soviet citizens, he saw eugenics as being integral to Nazism. Only when the American geneticist Hermann Muller was invited to join the prominent agricultural geneticist Nicolai Vavilov in Moscow in the 1930s did eugenics and genetics become linked in the Soviet Union. For the Marxist Muller, as much as for the Swedish social democratic Myrdals, a eugenic policy was an essential element in building the good collectivist society. His futurist text, *Out of the Night*, envisaged a genetic/eugenic society in which great men (above all Lenin), and absolutely no women, would be cloned, and the socialist population of the young USSR would qualitatively benefit.[7]

Unfortunately, with this Muller did his Soviet geneticist colleagues no favours; there was guilt by association for both Soviet genetics and Soviet geneticists. The class origin of the geneticists, in Vavilov's case bourgeois, added to their vulnerability. By 1940 Vavilov was imprisoned and genetics disgraced as being synonymous with eugenics. After the war, and following the triumph of Stalin's favoured agronomist Lysenko, geneticists with an impulse to survive rebranded themselves as radiation biologists concerned with the mutational effects of nuclear fallout following Hiroshima and Nagasaki – with the physicists aiding their survival strategy. Genetics was not to recover as a discipline in the USSR until well after Stalin's death and the subsequent fall of Lysenko.

Muller was luckier; tipped off by his colleagues, he left the Soviet Union just in time, returning to the US and going on to win a Nobel Prize for his work on mutations and the effects of radiation while still a committed eugenicist. His vision of positive eugenics was ironically realised in 1980 as 'The Hermann Muller repository for germinal choice' by the wealthy businessman Robert Graham. Graham's initial plan for his sperm bank was that it should be for the donated gametes of geniuses, with the Nobel Prize winner as the idealised donor. Muller's widow was outraged by the use of her husband's name and insisted it be removed. Ironically it was the right-wing ideologue of scientific racism, physicist Nobel Laureate William Shockley, who loudly advertised his donation, thus demonstrating his ignorance of the genetic uncertainties of inheritance. The sperm bank with its combination of androcentricity and fantasy genetics could not and did not last for more than a few years, and after the birth of some 230 children Graham's descendents closed the bank down and destroyed the records.

Muller's position was essentially that of liberal and Western Marxist biologists in the 1930s, who combined eugenics with anti-racism. The Seventh International Genetics Conference, originally planned for Moscow in 1937, was unable to take place as Stalin refused to allow eugenics to be discussed in order to avoid giving a platform to Nazi race theorists. US geneticists, much closer to race theory, did want to discuss its scientific foundations and its significance for eugenics. Faced with these insurmountable conflicts the plan for the conference collapsed. Two years later it was reconvened in Edinburgh, where a group including Haldane, Hogben, Huxley, Needham and Dobzhansky, with Muller as the key drafter, responded to the question 'how could the world's population be improved most effectively genetically?' by issuing what became known as 'the Geneticists' Manifesto'. Nothing could be achieved, they said, without major changes in social conditions providing equality of opportunity and removing race prejudices, which had no basis in genetics. Once this was achieved, along with effective birth control to enable women to contribute equally in work and culture, eugenic policies based on sound biological principles would enable the conscious direction of human evolution – though who exactly was to direct it was left unstated. The political commitment of these left and liberal scientists to introducing social measures prior to eugenics was indeed progressive, but they were blind to the possibility that birth control could be a mixed blessing, adding to women's freedom but also recruitable for eugenic ends determined by powerful others.

EUGENICS IN THE WELFARE STATE

Although the word eugenics was politically tainted after the Nazi death camps, it is important to slay an often-perpetuated soothing fiction. As a policy and practice eugenics neither originated nor died with the Nazis. Given the weight of historical scholarship mapping the eugenic programmes, particularly of the European welfare states still in place for some thirty years after the defeat of the Nazis, it should be unnecessary to speak of slaying this particular fiction, but the whole power of soothing fictions lies in their hydra-like reproductive capacities. Cut off their heads and they simply proliferate. Pointing out the widespread forms of state eugenics from the 1930s to the mid 1970s (with Slovakia still compulsorily sterilising Roma women today) is not to de-demonise the Nazis, but to position this particular and terrible episode in the historical context of mid-twentieth-century collectivism, within which, as we discussed in the Introduction, the well-being of the collective was more highly valued than that of the individual. There were public gains and private losses in that stance, just as there are in today's overvaluing of individual choice and undervaluing of public well-being.

An uncomfortable but fruitful way of thinking about National Socialism is thus as the unacceptable face of that great achievement of the twentieth century – the welfare state. Within the welfare states, eugenic policies sought not only to regulate the reproduction of those defined as biologically unfit but also to encourage the reproduction of the fit, and hence policies supporting the 'normal' family. Nazi Germany's positive eugenic policy was initially not that dissimilar, with public tribute paid to Aryan mothers with large healthy families. But as its policies became more extreme, Germany's positive eugenics took the form of breeding superman. Suitably blonde women from occupied countries were selected then raped by SS troops to bear pure Aryan children.

In Britain, positive eugenic incentives were rather more modest. As junior academics at London University in the 1960s, we both received £50 per year added to our salaries for each of our children. The explanation for this largesse was that William Beveridge – highly influential in the London University system when director of the London School of Economics – was a keen eugenicist and took the view that university teachers were self-evidently 'fit' and should be encouraged to have children. His envisaged university teacher was, it may be reasonably assumed, male, heterosexual, married and able-bodied, as indeed they mostly were. What few academic women there were, were almost entirely single and, like their male counterparts, white.

Rather more surprisingly, compulsory sterilisation continued until the mid 1970s in the Nordic countries, seen as necessary for the management of the nationalistic social democratic welfare state, just as the Myrdals had proposed at the time of its formation. Sterilising those who were 'unfit' to parent (in the double sense of having children and caring for them) reduced the potential burden on the state and enabled it to provide universal high quality welfare for the fit. Gunnar Broberg and Nils Roll-Hansen's *Eugenics and The Welfare State: Sterilisation policy in Denmark, Sweden, Norway and Finland* shows that these programmes were overwhelmingly directed against women, even though male sterilisation was easier and safer, while for women in a pre-antibiotic age it was both difficult and dangerous.[8] Scandinavian biomedical science in the service of the welfare state saw itself as entirely able to judge that certain women should be sterilised because they were 'unfit to mother' in two distinct senses: unfit to breed and unfit to care for their children. It was the latter which sustained eugenic polices, whether of compulsory sterilisation or sexual segregation, for so long. While the raw violence of compulsory sterilisation shocks, at least Scandinavian women lived in the community; unlike their incarcerated Dutch and British sisters they were not locked away like criminals.

The last case of state eugenics in Western Europe occurred in Cyprus, directed not against the learning disabled but as an attempt to reduce the incidence of thalassaemia. This genetic blood disease endemic to the Mediterranean was, in its harshest and most painful forms, widespread on the island, with around a quarter of the Greek Cypriot population being carriers. Archbishop Makarios, who had provided both political and religious leadership in the anti-colonial struggle against the British, became the first elected President in 1960, creating an unrivalled fusion of state and church powers. First came health promotion and education programmes designed to reduce the incidence of the disease, with more concrete preventive strategies becoming possible once a DNA test was available. Couples planning to marry with the blessing of the Orthodox Church were required to be tested for being carriers of thalassaemia, with the clear message that if the risk to any children was too great the marriage would not go through. Given the importance of a Church wedding in Greek Cypriot culture this was a powerful tool, and by 1986 the numbers of newborns homozygous (that is with two copies of the mutant gene, one from each parent) for the disease had dropped tenfold.[9] It is inconceivable that no pain and stigmatisation occurred as a consequence of what was effectively compulsory testing, but there was also widespread relief as the risk of the feared condition radically diminished.

The only comparable example of a quasi-statist eugenics was that of the New York rabbi Moshe Feinstein's response to the prevalence of the fatal genetic condition Tay-Sachs among the Ashkenazi Jewish community. Babies with the disease live in excruciating pain and die within a year. Feinstein persuaded the community that their young people should provide information about their carrier status to him, so that when courtship developed either partner or both could consult the rabbi as to whether marriage and children were a safe proposition. Trust in Feinstein with his concern to avoid stigmatisation meant that his project has been successful, but, like that of Cyprus, the case also raises problems for those hostile to state eugenics with all its authoritarianism. Meanwhile, the mass incarceration of African American men not only gives the lie to the *Myth of Black Progress*, as Becky Pettit points out, but as long-term sexual segregation also serves as covert eugenics.[10]

EUGENICS AND THE BIOLOGISTS

After 1945 in Europe and the US, the vocabulary of eugenics was slowly eased out of respectable scientific research, to be replaced by that of genetics and molecular biology. Davenport's Cold Spring Harbor Eugenics Record Office was closed in 1940; the kind of eugenics it represented had been scientifically discredited and the Office's many thousands of painstakingly filed records were discarded as valueless. Withdrawal of funding by the Carnegie Foundation and Davenport's retirement did their work. With the success of the double helix in 1953, Cold Spring Harbor found a new identity as the Laboratory of Molecular Biology, with Jim Watson becoming its director in 1968.

In Britain, the medical geneticist Lionel Penrose was appointed to the chair at the Galton Eugenics Laboratory at University College London in 1945, and ensured its renaming as the Department of Genetics. Penrose's own pioneering research on the aetiology of 'mental defect' among children was key in this renaming. He drew attention to the diversity of conditions which fell under the label 'mental defect', and hence the problem of identifying any clear-cut phenotype. The immense difficulty of establishing causation, disentangling the parts played by social and family environment and genes, helped dispel the idea that there were simple causal heritable factors. By showing that what is now called Down's syndrome was a developmental problem rather than a genetic 'reversion' to a 'less-evolved Mongoloid type' leading inexorably to mental defect, he provided the science which helped replace the stigmatising category of 'the Mongol' with that of a person with Down's syndrome.

The term Mongol harks back to a racial classification made two centuries ago by the German physician Johann Blumenbach, who divided humans into five distinct races based on skin colour. He called the 'highest type' Caucasian, arguing for what he considered the beauty of the white skin of the Georgian peoples of the Caucasus. Although Blumenbach's four other hierarchical racial categories have ended up in the dustbin of history, as both racist and genetically meaningless, the term Caucasian has stuck and is used extensively in the US to refer to people of European origin. While the guidelines issued by the Race, Ethnicity and Genetics working group of the American National Human Genetic Research Institute rightly emphasise the need for caution and precision in categorisation, they nevertheless fail to discuss 'Caucasian', a classification commonly used in US biomedical research. Given the hard political struggle to eradicate the racist categories left over from slavery it is difficult to understand why US geneticists still cling to Blumenbach. Perhaps if there was greater awareness that Russian racists still call people coming from the Caucasus 'blacks', they might feel a greater need to deal with the elephant in the genetics lab.

Despite Penrose's work, the biologists' enthusiasm for eugenics was slow to fade. Peter Medawar, the director of the National Institute for Medical Research, was still fretting away in his 1958 Reith Lectures for the BBC at the old eugenic question of whether the welfare state had interfered with natural selection, so that the quality of the UK national stock was deteriorating. He failed to see that the British practice of incarcerating learning disabled people – mostly women – was already part of an unspoken eugenic policy. Medawar was not displaying some freakish eugenicist aberration. In the 1960s, Richard Titmuss and his economist colleague Brian Abel-Smith proposed an unequivocally eugenicist child-allowance policy to the government of Mauritius. On the assumption that large poor Mauritian families were inhibiting economic growth, they proposed that child allowances be restricted to those families who did not exceed a predetermined number of children. Despite their commitment to defending British children from poverty, large poor Mauritian families were to be punished with even deeper poverty.[11] Titmuss and his colleagues' defence of universalism and collectivism did not extend to the families of a sometime British colony.

Such everyday eugenicism was slow to die. The language of the 1967 British Abortion Act is distinctly eugenic, indeed its successful movement through the legislative process rested on an unstated alliance between eugenicists and feminists. In practice it soon became abortion on demand for early pregnancies. The transformation was part and parcel of the change in women's consciousness – both in women wanting a termination and women working as counsellors, nurses or clinicians, and indeed in the consciousness of many men. This transformation of women's understanding of their gendered identity – a good deal less ambiguous than the universalising concept of personhood commonly deployed in current social theory – was not brought about by any scientific discourse but by the power of a near global feminist movement.

Even in 1968, that symbolic year of radical hope, eugenic enthusiasms were still being openly expressed:

> There should be tattooed on the forehead of every young person a symbol showing possession of the sickle cell gene or whatever other similar gene ... It is my opinion that legislation along this line, compulsory testing for defective genes before marriage, and some form of semi-public display of this possession, should be adopted.[12]

The painful thing is that the author of this stigmatising proposal was not some bizarre figure from the right, but a hero of the anti-war and alternative health movements, none other than the Nobel Laureate Linus Pauling. Taking Vitamin C in mega doses, whatever else it achieves – Pauling believed it protected against the common cold – clearly offers no protection against disability-hating thought.

Nor were Pauling and Medawar alone among biologists in their eugenic enthusiasms. At the BSSRS meeting in 1970 on the Social Impact of Modern Biology discussing the imminent prospect of IVF, the socialist immunologist John Humphrey spoke of 'getting rid of it', referring to the infanticide of impaired newborns.[13] To be fair, and perhaps because he was a father, he did express concern about the feelings of the mothers – unlike the childless, MIT cognitive psychologist Steven Pinker more recently. For much of the second half of the twentieth century biologists had worried lest the social provisions of the new welfare state enabled too many 'unfit' to survive, thus damaging the quality of the national stock and creating an economic and social burden. This anxiety was shared by the state but was understood as needing discreet management. For some biologists at the turn of the millennium, most vocally the geneticist Steve Jones, the worry was now that modern medical technology, by enabling the survival of the 'unfit', was preventing the winnowing effect of natural selection central to the Darwinian mechanism, thus bringing evolution to a halt.[14] But the systematic erosion of the welfare state had opened up the class differentials in both morbidity and mortality. Winnowing was continuing apace, and the role of medical technology in any of this was relatively trivial. Poverty does a much more effective job.

Racism has played a dysgenic role in British towns and cities with substantial Asian populations. The discussion has often focused on Bradford, with its large Muslim population and high incidence of babies born with some impairment. Though public health officials and geneticists have regularly pointed to the negative effects of cousin marriage in increasing the risk, such marriages have significantly increased over the years and with it the percentage of newborns with impairments. Health education campaigns have manifestly not worked as the pressures for cousin marriage outweigh the genetic risk. As immigration has been made increasingly difficult the community has turned to arranged marriages, often with a cousin from the home country. In this way a community which feels beleaguered by Islamophobia strengthens its numbers, as marriage brings an automatic right of residence. Islamophobia, it seems, can be dysgenic.

THE BIRTH OF NON-DIRECTIVE MEDICINE?

The historians' steady demythologising of the vision of 1945 as a moral watershed for biomedical ethics is not simply a debunking of a naive hope. Informed consent as a practice – rather than purely a matter of the publication of ethical reports and elevated discussion – was slow to appear. But for one medical specialism, 1945 was indeed an ethical watershed. Clinical genetics more than any other specialty had to confront the appalling revelations of what their German professional colleagues had been engaged in: the biomedical categorisation, compulsory sterilisation and finally extermination of the 'unfit'. Purging clinical genetics from the moral contamination of the Nazis' eugenic project was crucial. It was against this history that clinical genetics had to reconstruct itself. Partly it was about changing names – rebranding the profession, to use today's advertising language – but also and mainly about developing new ideas and practices for patients' informed consent. Counselling, spending time with patients, making sure that they understood the risk and that their decisions were freely taken, were to become the hallmarks of best clinical practice.

The non-directive nature of the genetic clinic all too often contrasts with the situation in the antenatal clinic. In the former, women and their partners at genetic risk but who want children are treated with sensitivity. In the latter, women have long been expected to comply with medical guidance on life-style regimes, give blood and urine samples, have amniocentesis and foetal scans – all for the good of the baby. The growth in molecular genetics has resulted in a proliferation of screening tests, but in Britain has not been accompanied by an adequate provision of counselling services. Screening tests to diagnose, for example, Down's syndrome, Cystic Fibrosis (CF) and Neural Tube Defect (NTD), to name three relatively common problems, will, it is argued, lead to fewer children being born with these impairments bringing relief to intending parents and also considerable savings in public expenditure. As tests to detect an increasing number of genetic risks become routine within antenatal care, they carry the risk that unless adequate counselling is offered women will feel powerless, as if on a conveyor belt and having little or no say in what happens to them or the foetus they carry. And the clinicians who care for them may be all too aware of this – one study of British obstetricians reported that 43 per cent were concerned at the lack of counselling available for their patients. Nor does antenatal testing, which claims to give reassurance, unambiguously provide it. The very act of testing throws into question the hitherto taken for granted 'normality' of the foetus. Without adequate social support such an experience can be more harming than calming.

For women in the rich countries pregnancies are no longer just pregnancies. As Barbara Katz Rothman's *The Tentative Pregnancy* describes, pregnancy has become a tentative state until the woman and her foetus have passed the scrutiny of genetic testing.[15] Rayna Rapp's decade-long study of antenatal care, *Testing Women Testing the Fetus*, was carried out when New York's antenatal care system was well funded and where an hour of counselling was available to every woman.[16] Rapp argues that the knowledge gained from the tests both increases women's control over their lives and at the same time exerts eugenic pressure. She speaks of these women as moral pioneers as they find their way through an increasingly dense minefield.

Even as DNA diagnostics expand, so the clinical context changes. This is nowhere clearer than in the case of CF, a mutation commonest among those of European descent where one in twenty-five people carry it. In the 1950s, most children born with CF died before the end of their first year of life, and that short life involved considerable respiratory distress. The hope in the heyday of gene-hunting was that if the CF gene could be identified then it would be possible to develop genetic therapy. The gene was identified in 1989 but the hype receded as hopes of gene therapies faded. Identifying the gene did, however, make possible a DNA test, enabling women to decide whether they wish to continue the pregnancy. Therapeutic management has improved and today, with adequate medical care, children born with CF can have a life expectancy of forty. If a social order can and does provide health care for all, then the options open to the woman and her family will be that much greater. Would she feel able to keep the baby if she has other mouths to feed, money is short, and her family has little or no access to health care? In this case the state does not have to coerce her – except by providing the information of genetic risk.

THE JANUS FACE OF CLINICAL GENETICS

Clinical genetics was and is caught in a major contradiction, intensified in the old welfare state countries of Europe as public health services are eroded. A service has to make both a moral and an economic argument that demonstrates its socio-economic value. This takes the form of cost–benefit calculations that turn on the collective eugenic advantages to be gained through providing clinical genetic services, particularly in the antenatal clinic. Thus, Janus-like, clinical genetics shows one face to the would-be mother and another to the state. For the most part it solves the contradiction in its position by making the assumption that the view of what constitutes a worthwhile life will be shared by clinician, the patient and the man she wishes to have children with. By making allowance for the reservations of a few women who will refuse abortion, clinical genetics argues that individual autonomy, informed choice and the state's wish to keep the numbers of children born with severe disabilities to a minimum will come together in a social and cultural consensus. Phillip Kitcher goes so far as to advocate this situation as a form of utopian eugenics.[17] But however attractive his appeal to an invisible hand may sound, it turns on the adequate provision of high quality counselling services. Utopian eugenics may be available for some but not for all.

The more sceptically minded observe that this double agenda produces an invisible norm determining the 'acceptable' number of children with impairments that can be born. Meeting the invisible norm satisfies the double demand of Janus – first that the pregnant woman is enabled to make an informed choice; second that the state's burden arising from the care needs of those born genetically impaired is minimised. Given the current political struggle being waged by the disability movement to insist that theirs is a life worth living, it is reasonable to view Kitcher's eugenic utopia as a shade Panglossian. The problem for at-risk women is to decide whether they are in the best of all possible worlds. Will their child be cherished as a shared responsibility of the family and the state, as in the golden age of Scandinavian welfare, or has the shrinking state – never that great so far as the disabled children of Britain have been concerned – chosen to give up its share of care? The combination of the state walking away from responsibility while providing genetic tests transfers responsibility to the would-be parents and above all to the would-be mother; genetics within a neoliberal political economy thus effectively helps construct a consumer eugenics.

Hardly surprisingly, the movement of disabled people is deeply suspicious about the proliferation of genetic testing and sees it as inherently eugenic.[18] People with learning disabilities have been at the forefront of these confrontations. In a culture that has for so long taken it for granted that a woman would not want to give birth to a child with Down's Syndrome, and would see screening as helpful, these challenges are both disturbing and long overdue. Defending women's reproductive freedom while refusing to subscribe to an automatic categorisation of the life of a person with Down's as not worth living is not easy. When the test can only indicate presence or absence and not the severity of the condition the argument becomes unsustainable, as is evident in the several debates between those who support these antenatal tests and learning disabled people arguing their rights to life itself. The belief that all would-be parents want a normal baby – as defined by medical doctors – is beginning to crumble. Hearing people uncomfortably learn to listen to the claims of deaf couples that they would much prefer a deaf child and would like to use pre-implantation genetic diagnosis (PGD) to secure their desire. Juxtaposed to the wishes of the parents is the child's right to an open future. This is new moral territory. Instead of the old medically defined normality simply reigning unquestioned, ethical arguments increasingly have to be developed and argued for democratically.

When Jim Watson was once asked what he thought about the eugenic implications of the Human Genome Project, he replied, 'well its no fun being around dumb people'. His remark was widely seen as offensive, but what is so very different, certainly so far as political activists with learning disabilities are concerned, between Watson's characteristic bluntness and the routinised offering of amniocentesis and chorionic villae tests to women over thirty attending the antenatal clinic? Both are expressions of the dominant cultural assumption that the presence of learning disabled people within the family and within society at large is problematic. Watson's comment at least has the merit of honesty; it is not evident that all those who found it offensive can claim the same.

While compulsory state eugenics has retreated, there is still plenty of eugenic pressure within neoliberal culture, economy and society. And where race is as central in culture and politics as it is in the US, the pressure can still centre on race. The African American sociologist Troy Duster – a leading figure in the ELSI programme of the Human Genome Project – carried out a study in the 1980s of the Californian introduction of genetic screening programmes for poor African American families. In the context of the grudging approach to welfare by California (to say nothing of its history of energetic compulsory sterilisation of African American women), funding the programme was resolved by cutting the material support these families had hitherto received. Duster pointed to the contrasting approach in the UK, where similar DNA screening programmes had been discussed but were rejected on cost–benefit grounds. In the context of the UK's more generous welfare provision, even in Thatcher's Britain, social rationality had more space to thrive and eugenics less. The Californian screening programme, by contrast, became a back door for eugenics. Duster concludes:

> Once again whether this new genetic knowledge is an advantage or a cross depends only partly on how the genes are arranged. It depends as well where one is located in the social order.[19]

So much for utopian eugenics.

SLIDING INTO CONSUMER EUGENICS

Human embryo screening arrived with IVF. Initially this was done by 'eyeballing' the embryos and deciding which of them met the biologist's tacit understanding of the healthy normal egg. Pre-implantation genetic diagnosis goes a step further by taking a biopsy of the embryos for genetic screening. PGD has been regarded with unease in Europe, above all in Germany with its hard earned understanding of eugenics. In Britain PGD is regulated so as to be available only where would-be parents are at risk of having a profoundly disabled child. At the heart of the problem is the decision as to what constitutes such disability. PGD is potentially available for many single-gene disorders. However, as with Cystic Fibrosis many of these can be modified and managed, even though not 'cured'. Thus the successful introduction of statins plus life-style management has seen a substantial reduction of morbidity and mortality for inherited lipidaemias such as Familial Hypercholesterolaemia. Should PGD be deployed in such relatively manageable situations, allowing these less than perfect embryos with their still reduced expectation of life to survive, or should they be rejected? Is PGD helping to promote the ideology that every parent has the 'right' to a perfect child? Despite the British claims to have well regulated these issues, they seem to be offering modest protection against creeping consumer eugenicism. The permission to use PGD granted to one couple at risk of having a child with a cleft lip and palate, when surgical repair is possible, seems to be pushing the limits.

With PGD come new ethical dilemmas – the prospect of 'designer babies', and 'saviour siblings'. In the former, the selection is made not against some profoundly disabling genetic condition but towards enhancing some desired characteristic, from sex to hair and eye colour and, speculatively, beauty or intelligence. While these remain at the level of fantasy, saviour siblings are already among us. These are children born following PGD and embryo selection to provide a tissue match for older sibs suffering from some serious genetic disorder. Bioethical debates within UNESCO and similar elevated settings anticipated such possible technical developments, and rejected them as ethically abhorrent, as turning the genetically selected child into a mere instrument and also abusing her right to bodily integrity, for there was no possibility of eliciting her informed consent on whether to give or withhold her tissue. But this kind of opposition from bioethics frequently melts like snow in summer in the face of a family tragedy playing itself out in the media.

The first such story to hit the headlines, in 2000, was that of the Nash family in the US. The saga of the Nashes and their now two children has served as a microcosm of current genetic testing/eugenic anxieties. Have the well-heeled Nashes opened the door to the commodification of children and the eugenicism of consumer society, or do they simply represent an expression of parental love attempting to muddle its way through our biotechnologically advanced times as humanely as possible? For Molly Nash, a six-year-old with the fatal genetic disorder Fanconi's anaemia, the best chance of life was a transfusion of umbilical cord blood from an unaffected sibling. The Nash's second baby, Adam, was chosen through PGD to be free from Molly's life-threatening genetic condition and also to provide a life-saving resource for his sibling. The sense of horror that a child could be brought into the world as a compulsory donor to save the life of an already existing child competed with the claim that this was the only way the existing child could be saved, that human beings had children for a number of complicated reasons, and that, as the Nashes wanted more than one child, a baby born as a result of these procedures would be loved anyway. The couple and their children were extensively interviewed and photographed in the process, thereby normalising and personalising a morally controversial innovation.

The Nash case was swiftly followed by another, this time in Britain. Raj and Shahana Hashmi, whose son was suffering from the blood disorder beta-thalassaemia major, campaigned publicly for permission to use PGD to select an embryo free of the condition whose cord blood could be used as it had been in the Nash case. Given that thalassaemia is genetically transmitted, some were troubled that the Hashmis had not checked out their own carrier status, which might well have ensured that their first son was free of the condition, but the media once again gave prominence to pictures of the happy, articulate and good-looking couple with their very ill child, and the HFEA obliged with a licence.

Such cases exemplify the development of a consensus that parents in these situations will love both children equally, that the older sib will be saved and the younger content with their unnatural selection, and indeed that this is a happy ethical story for all involved. Those not quite able to sign up to this heart-warming tale – who worry about what propels parents to advertise their most intimate distress on the media and who have a sense of the contingency of family relationships – are left with the dour adage that bad law tends to be made from hard cases. The arrival of such babies demands serious moral reflection on the part of civil society collectively, and not just from the mediatrics of the professional op-ed writers.

The difficulty of holding any line against selecting saviour siblings – or, to put it most brutally, against growing spare-part siblings – was nowhere better demonstrated than in the Whitaker case. Because in this case the existing child's life-threatening mutation was not genetically transmitted, the HFEA refused assent. In an age of reproductive tourism, the Whitakers flew to the weakly regulated milieu of the US for their PGD and IVF. The fact that such tourism is possible does not, however, mean that nation states should give up on trying to regulate, and instead follow the market libertarian position of letting everything go in response to the so-called technological imperative. Rather few lawyers imagine that making murder illegal and punishing murderers actually stops murder; it does however limit its routine practice as a strategy for conflict resolution.

In his prize-winning novel *Never Let Me Go*, published just three years after the Hashmi story made the headlines, Kazuo Ishiguro turns the ethical issues raised by the PGD-selected foetus into an imagined nightmare in which children are cloned to provide replacement parts for others. These spare-part children are aware that at some time they will be called on to be donors to provide tissue-replacement organs for more human others. Their entire socialisation is directed towards them accepting this future, their only comfort being their love for one another. Ishiguro's dystopia of the breeding of spare-part children, along with that of Margaret Attwood's *The Handmaid's Tale* and *Oryx and Crake*, offer warnings of a dystopic eugenic future it would be a mistake to discount.

STANDING ROOM ONLY: SEVEN BILLION PEOPLE

Ever since Malthus, 'Standing Room Only' has been the leitmotif of the population growth panic-makers – except when they are fretting about the decline in the birth rate and the increasing 'burden' of an ageing population. Malthusian inevitability has been massively circumvented by human ingenuity that has so far avoided his predictions for the consequences of population growth. Those within the specialist world of demography on the whole follow the ups and downs of life expectancy and family size rather more calmly, and tend to be cautious about projecting future birth rates. The moral panics tended to be generated by the biologists, who have and still do swing about quite alarmingly. Thus in 1934 the mathematical biologist Edith Charles, subsequently an adviser to the WHO, published *The Twilight of Parenthood: A Biological Study of the Decline of Population Growth*. At the time this was a handy weapon to deploy in her and her husband's (Lancelot Hogben) opposition to eugenics, then in ideological full flood. Population control policies, something of a euphemism for eugenics, were at their height in the 1960s and '70s, seen by the rich West as saving the planet by reducing the populations in the poor and most populous countries. The Club of Rome, Prince Philip – himself the father of four – and ecologist Paul Ehrlich sounded the alarm. The barbarity of the West's eugenic ideology was nowhere better expressed than in Ehrlich's 1977 book *The Population Bomb*.[20] 'Cancer', he wrote, 'is an uncontrollable multiplication of cells; the population explosion is the uncontrollable multiplication of people.' With the birth of the seven billionth person on 31 October 2011, 'Standing Room Only' is bidding to make a comeback. From the magazine *Prospect* to Nobelist John Sulston, concern about population growth is back on the cultural and possibly political agenda.

The solution echoes that of Malthus, according to the most fundamentalist of today's population eugenicists. Disturbingly, these include the chair of President Obama's science and technology advisory committee, the environmental scientist John Holdren, for whom ecological disaster is inevitable unless the world's population – including America's – is reduced by a global authority with the power to overturn human rights. While preferring more voluntary approaches, the book Holdren co-authored with the Ehrlichs in the 1970s envisages the use of compulsory sterilisation, compulsory abortion, and anti-fertility drugs in the water supply, with only those with a licence to have a child being given unadulterated water. They add: 'if the population control measures are not initiated immediately, and effectively, all the technology man can bring to bear will not fend off the misery to come'.[21]

The populous post-colonial countries soon recognised the Malthusian agenda of the environmentalists as a form of racist eugenics, knowing that in conditions of poverty large families make better sense in the short term. It was not until the feminist movement of the '70s, which called for social development and above all womens' reproductive freedom, that the politics of population control were modified. Over time, what had been population control increasingly became a birth control movement for women, part of a political struggle to enhance their status, in which education was integral. In this context, where women are more valued, better educated and have the space to make their own decisions, they do have fewer children. Africa is the exception; there the four horsemen of Malthus's apocalypse – war, famine, pestilence and death (the latter in the form of AIDS and malaria) – ride over much of the continent.

India and post-revolutionary China have long been conscious of the challenge of feeding their vast populations and the need for limiting the size of families. In India, as Amartya Sen has observed, once the British had left, there were no more famines, and family size was gradually reduced through birth-control programmes. While this approach did slow down population growth, it was not fast enough for President Indira Ghandi. With appalling political misjudgement, in 1975 she gave her son Sanjay the responsibility for driving through a policy of the compulsory sterilisation of 8 million peasants, resulting in massive protests. The policy was only terminated after Sanjay's assassination in 1980. However, to focus exclusively on the state, rather than on the eugenic practices of the patriarchal Indian family, would be to ignore the formidable contribution of DNA testing, which has made antenatal sex selection possible and hence intensified the pressure on women to abort female foetuses. While still practised by poorer families, female infanticide has become technologically obsolete for the better-off. In the 'normal' sex ratio, around 105 boys are born for every 100 girls; but in India by 2010 there were 914 girls aged between 0 and 6 for every 1,000 boys. India has one of the most imbalanced gender ratios in the world. Sex selection, infanticide, discrimination in access to food and health care, even murder, are all part of this domestic femicide. The government legislates but with little effect; until women have equal status it is difficult to believe that enforcement and conviction will be much more successful than most countries' rape laws.

Unlike India – China, which since 1978 has systematically employed state eugenicist measures to secure population-growth management – still suffered famine as a consequence of Mao's disastrous economic policies. The one-child family policy has reduced family size, slowing down population growth, but has, as in India, resulted in a severe gender imbalance. Official Chinese statistics currently report that for every 100 females nationally there are 119 males. In the most severely imbalanced province the ratio is 100:137. Despite the Communist Party's ostensible commitment to gender equality it has presided over the compulsory sterilisation of some 20 million Chinese women. From within the patriarchal family there has been steady pressure on women to have sons. In the past, female infanticide met the need. Illegal DNA testing offers antenatal sex diagnosis and the abortion of female foetuses. Failing abortion, there is still female infanticide and increasingly giving up girl babies for both national and international adoption. There are claims that there have been compulsory abortions, and others that children have been taken from their parents when the family size has been exceeded. There is some more cheering evidence that as urbanisation proceeds, and mechanisation replaces heavy manual labour in the countryside, girls are being increasingly valued, and the gender ratio is slowly returning to something closer to equality. In the meantime, today's only children are loved and stuffed in equal measure, with the result that, as in the West, obesity and diabetes are becoming pandemic.

IMAGINARIES OF THE NEW EUGENICS

Is there a dividing line between therapy and enhancement, and is there any difference in principle between genetic enhancement and any other form? And how can either therapeutic or genetic enhancement be sensibly discussed when we are in the midst of a veritable storm of body enhancements provided by an expanding industry of cosmetic surgery? Nose jobs, breast enlargements, liposuction, face-lifts, botox, etc., have become routinised for women, and, as the zombie like-face of Silvio Berlusconi advertises, increasingly for men. Those in search of the beautiful face, the beautiful body, are encouraged to believe that the surgeon's scalpel can provide it. New standards for the appropriate jaw line or the desirable breast size pressure new conformities, while globalisation offers bargain prices for cosmetic tourists.

Therapy implies restoring some impaired bodily function; enhancement is about adding to an assumed already 'normal' well-functioning body. Though in an intensely scientific culture, where biomedicine has both taken and been given the authority to name nature, who gets to define what counts as a 'normal' body is less of a puzzle. Growth hormones have been prescribed for shorter children even though being short is not a disease. As the socially successful tend to be taller, ambitious parents in the US have been paying anything up to £100,000 for a course of treatment for their shorter children. Similarly some parents of children with Down's in the US have turned to cosmetic surgeons to modify their child's eyelids, ostensibly to protect the child from social rejection. Yet this leaves in place the hostility of the so-called normal to the child they stigmatise as sub-normal. Surgeons have long assigned a gender to babies whose genitals do not correspond to the biomedical binary of female and male. But while the surgeons have meant well for many of the subjects, feminist biologists Anne Fausto-Sterling[22] and Rebecca Jordan-Young[23] (both also science studies scholars) have shown that the outcomes are frequently negative. Regular internal examinations have been reported by both the children and their mothers to be sexually humiliating, and repeated surgery painful. The rise of the identity movement of intersex people has challenged the right of biomedicine to construct sexual identity in line with its binary concept, its scalpels and its drugs.

Such bodily enhancement is not, of course, eugenics, except in so far as it seeks to make the enhanced appear to have been wellborn, or even better than wellborn. But the new genomics has raised once again the prospect of enhancement by genetic engineering (germ-line therapy) and artificial selection held out by Muller and others in the 1930s. Even while the Human Genome Project was nothing more than a pipe dream, the molecular biologist Robert Sinsheimer, whose early enthusiasm for the sequencing project we described in Chapter 1, showed himself to be not a mere culler of the unfit like Pauling, but rather a new Muller for a new time – prophesying nothing less than a new eugenics which could perfect humankind. Thus in 1969 he declared that

> A new eugenics has arisen, based upon the dramatic increase in our understanding of the biochemistry of heredity and on our comprehension of the craft and means of evolution … The old eugenics would have required a continual selection for breeding of the fit and a culling of the unfit. The new eugenics would permit in principle the conversion of all the unfit to the highest genetic level. The old eugenics was limited to a numerical enhancement of the best of our existing gene pool. The horizons of the new eugenics are in principle boundless for we should have the potential to create new genes and new qualities yet undreamed … Indeed, this concept marks a turning point in the whole evolution of life. For the first time in all time, a living creature understands its origin and can undertake to design its future. Even in the ancient myths man was constrained by his essence. He could not rise above his nature to chart his destiny.

> Today we can envision that chance and its dark companion of awesome choice and responsibility.[24]

Writing almost thirty years later, in 1997, US molecular biologist Lee Silver's imaginary of *Remaking Eden: Cloning and Beyond in a Brave New World* rejects Sinsheimer's utopia.[25] Where for Sinsheimer molecular genetics will enable a perfected germ-line for the entire human species, bypassing the old eugenic anxiety about the increasing numbers of the unfit, Silver's vision is dystopic. His remade Eden warns that without adequate public policy, molecular genetics might well compound current inequalities. As the biotechnology develops the capacity to genetically engineer foetuses, it will not be located in Sinsheimer's fantasy world with the molecular biologist as God busy on the eighth day, but in the weakly regulated and hyper-marketised economy of the United States. Silver's imaginary is projected technologically but always in the context of an unchanged neoliberal social formation. He does not share the optimism of the feminist science fictions of the '80s from Ursula LeGuin to Marge Piercy, who envisioned a future in which the new reproductive technologies were part and parcel of a new social formation in which gender no longer deformed the relations between women and men. Indeed, for both writers sex/gender itself was blurred: LeGuin refused the possessive pronouns of his and her, and in the imaginary of Piercy a bearded man breastfeeds a baby.

But the 1980s and utopian feminism are distant memories. In Silver's more jaundiced view those who can afford to specify the physical and mental features of their future children will do so, much in the same way that the well heeled have always bought social privilege. The best private nurseries and schools, the coaching, the acquisition of accomplishments, the elite university; anything that can secure a competitive edge for their offspring will be supplemented or supplanted by gene therapy, a consumer choice available to all who can afford it, for reasons expressed most succinctly nearly a century ago at the crest of the eugenic wave of the 1920s, when American psychologist Edward Thorndike wrote: 'In the actual race of life, which is not to get ahead, but to get ahead of somebody, the chief determining factor is heredity.'[26] In Silver's Eden, the new technology would enhance and speed up this social process with the result that the population would eventually split into separate species – gene-rich and gene-poor. For him, human cloning was the key reproductive technology. Within six years of *Remaking Eden* the first mammal was cloned, involving no fewer than three mothers; the provider of the egg, the provider of the nucleus to be transferred into the egg, and the third to carry the embryonic Dolly. The realisation of Muller's dream and of Silver's concern was almost there.

The Hollywood film *Gattaca*, which came out in 1997, the same year as Silver's book, constructs another Eden, one in which only the gene-rich, selected this time by PGD and called 'valids', are qualified for key jobs. The hero, though not genetically up to snuff (and classified as 'in-valid'), manages by sheer John Wayneism – a stolen identity, tough training and surgical enhancement – to outwit his genetic destiny. True grit trumps genes. Something of a reversal of the growing culture of genetic determinism fostered by the hype of the HGP and the daily claims to have identified a gene for almost everything.

Silver recognises the advantage the rich have in both their economic and cultural capital, and very likely in the biological capital that comes with well-functioning bodies. But his ethical and political concern to inhibit the growth of class inequality is not shared by the British bioethicist John Harris, who as a market libertarian argues that if a genetic technology which increased intelligence or longevity was possible, it would not be so qualitatively different as to raise fundamentally new ethical problems.[27] Furthermore, he insists, there is no fundamental ethical difference between employing drugs or gene therapy to improve a child's exam performance or athletic ability and the private coaching which achieves the same ends for the children of today's wealthy. Both are attempts to direct a child's future, and both can be seen as forms of enhancement. Thus genetic enhancement, framed within neoliberal individualism, adds a new variant of positive eugenics to the Matthew principle of 'to him that hath, more shall be given'.

Harris's position is an unblushing defence of possessive individualism. To the already huge privilege of wealth and influence that rich parents provide their offspring, he wishes to add any potential enhancement these parents can choose and that the biotechnosciences can offer. Michael Sandel, a former member of the US President's Bioethics Council, comes to the diametrically opposite view, elegantly arguing the case against perfection.[28] Canadian bioethicists Baylis and Robert are unusual in that while hostile to capital's relentless drive for innovation, including that of the biotechnosciences, their opposition collapses with the conclusion that genetic enhancement is inevitable.[29] Whatever technical problems and regulative measures there might be, attempts at genetic enhancement are inevitable somewhere, most likely in the softly regulated environment of the so-called 'Wild East'. Such critics forget that while pessimism of the intellect is indeed necessary, so too is optimism of the will.

Philosopher Allen Buchanan, even while enthusiastically endorsing this prospect of radical enhancement, raises the problem of distributive justice. He asks how we can ensure that the benefits of enhancement are spread more equitably and do not just privilege the wealthy. Buchanan proposes a 'Global Institute for Justice in Innovation' to achieve this goal, which would, in the hands of a council of scientists, ensure the diffusion of enhancement technologies beyond wealthy individuals and societies on the model of trickle-down economics.[30] But in today's ruthlessly neoliberal world dominated by global capital, whether that of the US today or China tomorrow, reinventing Plato's ideal form of governance by the women and men of gold (the scientists in this case) beggars belief. Optimism is needed but not of the broken-wheel variety.

It has, however, been outside the life sciences that such technologically determinist imaginaries have flowered most rampantly. Just as in the 1980s the Repository for Germinal Choice attracted wealthy funders, though based on flakey genetics, so too today does the mushroom-like growth of transhumanist institutes and research programmes. In Oxford, the Future of Humanity Institute directed by philosopher Nick Bostrom, an advocate of 'radical enhancement', is financed by anonymous 'visionary philanthropists' dreaming of a positive eugenics for the twenty-first century. Among the most prominent of the new foundations is Aubrey de Grey's SENS (Strategies for Engineered Negligible Senescence), based in California and funded by Peter Thiel, the ex-CEO of the online payments system PayPal. Where two decades ago those wishing for immortality sought to have their heads deep frozen after death, to be thawed when the technology for revival became available some time in the future (so-called cryonics), today de Grey's vision bypasses freezing in favour of genetically engineered longevity, even immortality. Buying your way out of death has become part of the American dream.

Perhaps the most flamboyant of the transhumanists is the American entrepreneur and IT innovator Ray Kurzweil.[31] His imaginary sees rapid advances in informatics and biomedicine as inexorably driving human evolution, arriving within the next half century at what he calls 'the singularity' at which the combined powers of the new technosciences will come together at a nodal point and a transformed post-human biocybernetic species will emerge. With this Kurzweil realises Darwin's vision in which 'all corporeal and mental endowments will tend to progress towards perfection'. But where Darwin's teleological moment was a passing theoretical lapse, for Kurzweil progress to his dream of evolutionary perfection automatically follows the technosciences of life.

Deeply embedded in such transhumanist writings is the possessive individualism of the neoliberal economy, where the mantra is choice and the consumer is king or queen. The imaginaries conjured up by these bioethicists has much in common with those proposed by Galton, namely the desirability and feasibility of improving the human stock. The fact that it was only after the birth of the conjoint twins eugenics/genetics that state eugenic policies were introduced testifies to the power of the imaginary. Likewise, the fact that many geneticists were conscious that genetics could not sustain such biofantasies only emphasises their ideological commitment to the eugenic project. Their science served and was part of their ideology.

The excitable visions of the transhumanists amplified by a largely uncritical media serve to make any step, however small, towards the genetic bioengineering of humans appear reasonable and acceptable, as the sequential concessions of the HFEA and the HGC have shown. The judicial review's conclusion that the doctors yielding to parental pressure for PGD and abortion was permissible in the case of cleft lip and palate is but one example of the ethical salami-slicing that the alliance of the state, bioethicists and gung-ho biomedical technophiles is making possible. But the imaginaries of the molecular biologists have not been confined to consumerist projects of human perfectability, and with the funding of the state and venture capital they have taken what has been seen as the next step for biomedicine, namely putting clinical practice onto a firm genetic footing with the development of large-scale population DNA biobanks. These, when coupled to electronic health records, offer joined-up clinical care to clinicians and their patients while providing the state with increasing powers of surveillance. It is to this that we now turn.

5

*The North Atlantic Bubble**

The transition from the old collectivism to the new individualism characteristic of a neoliberal economy disturbed many assumptions about the relationship between citizens and the health-care system, above all in the context of the old welfare states. Once a source of pride for the social democratic vision, since the end of the long post-war boom the welfare states have been under increasing ideological pressure and subjected to severe financial cuts. Into this unsettled arena came the proposal for the first large-scale DNA biobank from deCode, a company owned and registered in the US but based in Iceland.[1] It was the height of the new biotech and dotcom boom in the mid 1990s, and venture capitalists – above all those in the US – were willing to risk large sums on biotechnology projects as the potential profits could be vast. Such biobanks were to become part of the new drive towards the commodification of the human body in the form of strings of DNA sequences – a drive integral to the promises of personalised medicine through advances in genomics.

'Iceland sells its people's genome' was how the media reported the arrival of deCode in 1998, giving rise to fierce public debates over the proposal to sell the entire population's medical records, including their DNA, to a private company. Though most Icelanders welcomed the proposed DNA biobank, a vocal minority spoke of it as the arrival of a brave new world in which there could be no privacy. The central figure in this saga was the charismatic Icelandic

* This chapter draws on Hilary's 1999 ethnographic study of the issues surrounding deCode as the first DNA biobank; all references unless specifically cited here are given in H. Rose, Endnote 1.

neurologist, Harvard Professor Kari Stefansson, deCode's CEO. But the media's avid reporting of what was for them and the public an extraordinary innovation, claiming to offer a route to identifying genes associated with many hitherto incurable diseases, masked the fact that such DNA biobanks had been quietly under discussion by insiders for several years as the HGP rolled on to completion. For this network of stakeholders – molecular biologists, Big Pharma, venture capital, and the state – biobanks were understood as a means to realise the biomedical benefits and the potential for wealth creation offered by genomics within a global market.

DNA biobanking puts individuals and their predicted future health status at the core of personalised medicine. Nothing could be further from traditional public health research, with its brief of protecting, and better still improving, the collective health of populations. The foundational public health story is John Snow's identification of the water pump in Soho, in the heart of London, as the source of the raging local cholera infection of the 1850s. He showed that removing the pump handle stopped the epidemic, and with this the disease was brought under control. This kind of intervention on behalf of the health of the collective is the antithesis of the new individualised predictive medicine. This seeks gene sequences in the individual that point to disease susceptibility; for example, to lung cancer – a matter of considerable interest to the tobacco industry. Reeling from the success of public health anti-smoking campaigns in the rich West, the industry sees the commercial benefits of genetically targeted marketing.

If the identification of genes associated with disease is difficult in the case of single-gene disorders, it is even more difficult in the case of disorders where more than one gene may have an influence. The problem – which will reverberate throughout our discussion of biobanks – is that if diseases are not inherited in a Mendelian way but involve many genes, each having only a small effect, interacting both with each other and with environmental variables, then how large a population must be surveyed in order to identify them? Further, if many genes are implicated, each contributing only a small amount to the risk of disease, can any useful conclusions for personalised medicine be drawn? In short, what constitutes a 'good population' for useful information to be derived from a biobank?

A central problem for all health-care systems is the rational management of the fast increasing drug bill together with the rising costs of treating those patients who suffer serious adverse effects from prescription drugs. In the NHS alone, such patients occupy some 1,000 hospital beds and cost an estimated £466 million a year. DNA biobanking offers to provide an entirely new approach through personal gene-based prescribing. This would supersede the approach of the UK National Institute for Clinical Excellence (NICE), which carries out meta-analyses of clinical trials, calculates the average quality adjusted life years (QALYs) to be expected from a new drug, and on this basis issues guidelines to the NHS as to if, when, and to whom they should be prescribed. The recommendations frequently come under attack from both patients hoping to extend their lives and from the doctors caring for them – most often in cancer care. The problem is that people's reactions to drugs are individual. The scientific limitations of NICE's advice mean that some patients who could benefit from a drug do not receive it. For example, it has long been known that some patients find the pain-killer codeine ineffective. Genetics provides both explanation and therapeutic guidance: 10 per cent of people of European stock lack a key enzyme, CYP2DG, that metabolises codeine. DNA biobanking and personal gene-based prescribing could remove the doubt.

Clearly, if simple and inexpensive tests were available to identify subsets of patients as either responsive or unresponsive to a drug, and treatments could then be tailored accordingly, it would offer better value for money to the health service and less discomfort and danger to patients – a win-win situation. This was the prospect that the deCode biobank held out. By combining medical records from the Health Sector Database (HSD) with the genealogies and DNA analyses from the biobank, disease-associated gene variants could be discovered, and then, hopefully, relevant existing drugs prescribed or new ones invented. In the context of rising health-care costs, it also offered the government a potential means of managing resources more effectively. DeCode's proposal thus shrewdly positioned itself to be attractive to the key players: venture capital, Big Pharma, national health systems and their marketised counterparts such as the US Health Maintenance Organisations, and the insurance industry. All were potential buyers of this new commodity – bioinformation – although their purposes are very different.

FUNDING AND BRANDING DECODE

The conflict over the Icelandic database broke out in 1998, but its origins go back to the summer of 1994. Stefansson and his colleague Jeff Gulcher were in Iceland to collaborate on a genetic study of multiple sclerosis (MS) with local neurologist and MS specialist John Benedikz. In what the Harvard researchers spoke of as 'helicopter science', they flew in during the summer, secured as many samples as possible from patients and their families, and returned to the US to do the lab work. For Stefansson, MS was but a starting point. His ambitions and vision were much wider than simply searching for the genetics of one disease. Although not a geneticist by training, he was the first biomedical researcher both to see the potential significance of the Icelandic genome to the genetics of what researchers speak of as common or complex diseases, and also how to bring together the potentially shared interests of the state and of venture capital. The diseases in question – above all heart and cancer – are common because they are the major killers, and they are complex because their causes are multifactorial. MS is a good example of the type of complex disease which molecular population genetics proposed to elucidate and develop new and effective drugs. Having spent two decades in the entrepreneurial culture first of Chicago then of Harvard (there is a marvellous quote in a *New Yorker* article where he explains how he had not worked at Milton Friedman's university for nothing), Stefansson was uniquely well placed to understand the research and commercial possibilities offered by the small, well-educated, rich and relatively isolated population of Iceland.

It was out of this vision that deCode was born, as a US-owned biotechnology company physically and ideologically located in

Iceland. From its inception, Stefansson had two very different, though synergistic, objectives. The first was to establish a commercial laboratory to carry out biomedical research in Iceland. This, like any other biotechnology company working on human genetics, would seek to collaborate with clinicians interested in specific diseases and, either alone or with other pharmaceutical companies, develop new DNA diagnostic tests and drugs. The second and highly ambitious objective was to integrate information linking the genetic profiles and health records of the entire population of Iceland – some 275,000 at the time. This integrated record was to form the Icelandic Health Sector Database, a resource for both the Icelandic state and deCode alike. Iceland's size, high-quality universal health care, medical records dating back to 1915, supposed genetic homogeneity, and large well-documented tissue bank serving as a potential repository for much of the nation's genetic record, plus the presence of the genealogies, were together seen as offering uniquely favourable conditions for turning the prospect of personalised medicine into a viable commercial project. In many of the discussions among social scientists about deCode these two objectives are conflated. Successes belonging to the biotechnology company have been attributed to deCode and understood as derived from the much more contentious proposed total population biobank. DeCode's offer was to collect and computerise the medical records of the entire population at its own expense and also offer an annual substantial fee to the government. This HSD was to become, both in the media and in much of the academic debate, virtually synonymous with the entire deCode project; its other entirely conventional gene-hunting objective was ignored.

Stefansson raised the funds to establish his research company – $12 million in single-dollar shares – over a period of a few months. The new firm, deCode, was registered in August 1996 in Delaware, long recognised as one of the most lightly regulated states in the US. *Red Herring*, a US magazine that seeks to bring together venture capital and would-be scientific entrepreneurs, soon listed Stefansson among its 'entrepreneurs of the year'. This did him no harm in either the US or Iceland. The initial venture capital firms were US based, though one had a UK partner, Vanguard Medica. Apart from the British Nobel Prize winner Sir John Vane, representing Vanguard, and Stefansson himself as CEO, there were initially no biomedical researchers on the board. Stefansson also recruited Vigdis Finnbogadottir, the former President of Iceland. DeCode's strongly national identity was thus constructed with some ingenuity. Delaware, so commercially important to the company, was elegantly tucked away behind the Icelandic facade.

Unlike the rest of Europe – increasingly concerned by the new risks posed by science and technology, most dramatically those arising from genetically manipulated organisms – Iceland's successful management of its geothermal energy resources together with its newly entrepreneurial culture were welcoming. Stefansson tapped into an old pre-risk rhetoric of 'science is progress; it cannot and must not be stopped'. Iceland's culture thus provided this first large-scale DNA biobank a peculiarly friendly niche. For deCode and genomics the 'good population' offered by Iceland lay in the story of its settlement, its smallness of size, and what was endlessly claimed without qualification, as the genetic homogeneity of the population. Geneticists have always been attracted to research social groups for whom genealogies are culturally important, whether Mormons or Icelanders. For the Icelanders genealogies have been a cultural passion since the original settlement a thousand years ago, providing a narrative of both personal and collective identity. There are few surnames, each indicating something like a clan, with dottir or son tagged on to indicate gender; consequently the first name is the most commonly used. The genealogies of these clans had already begun to be computerised as, following the fallout drifting over Iceland from the bombing of Hiroshima in 1945, the US Atomic Energy Commission had funded genetic research at the University of Iceland. DeCode built on this research and was soon able to announce that it had some 600,000 computerised genealogies.

In the deCode Corporate Summary of 1998, the original Viking settlers and their Celtic slaves who settled around AD 930 were socially upgraded to become 'Norwegian nobility and Irish slaves'. This narrative of homogeneity, of genetic nationalism, finds a popular echo not only in the deCode literature, but also in the symbolic representation of the nation, whether in the state-commissioned sculptures by the harbour or in the tourist-shop kitsch. The Viking helmet and the long ship predominate. The Celtic slaves – women acquired, often by abduction, for sex and breeding – have no place in this highly masculine symbolic self-representation of the nation. Despite deCode's claims, serology, the old technique of population genetics, indicated that Celtic genes are present in as much as 52 per cent of the population. Nonetheless, the zealous stressing of homogeneity in deCode's marketing successfully masked the complexity.

In the media furore over the database, stereotypical representations of an island people descended solely from 'blue-eyed blond Vikings' frequently spun out of control. In remarkably racist language, the Icelanders were described as 'a nation of clones' because 'everyone in Iceland is related to everyone else … all of them are descended from the same few Vikings…'. However it is difficult to believe that foreign gametes have not joined the local population in the usual way. Reykjavik is after all a significant seaport. There are east European migrant workers, and, as is common in many Western countries, there is also a growing number of internet brides – mostly from the Philippines. Like their Nordic neighbours, Icelanders also adopt children from the South. When US media commentary on the HSD began to invoke the stereotype as if it had a robust purchase on reality, the Icelandic Ambassador to the US intervened with a letter to the *Washington Post* to remind the readership of the population's diverse inheritance. Not only did he gently tease one of the journalists for suggesting that Icelanders were all blond and blue-eyed; he also recounted the story of how Hitler was so convinced that securing Iceland would be something like securing Valhalla that the German representative on the island found it necessary to write a regretful letter explaining that the Icelanders were rather more of a mixed bunch.

SPEED, THE MAGIC INGREDIENT

Stefansson returned to Iceland in 1997 and by November that same year had established deCode as a commercial research lab in Reykjavik. DeCode had soon spent more on research than the Icelandic government's annual research budget – some $65 million. At its peak it employed more than 400 staff of whom ninety had either a PhD or Master's degree – an enormous laboratory for Iceland but still small compared with the giant companies (Genentech in the US has over 11,000 employees). For Stefansson speed was of the essence. Speed first in getting the political and venture capital support in place, then in getting the biotechnology company established, and lastly in getting the Health Sector Database up and running under exclusive control. Speed was the magic ingredient that would provide the competitive edge for a new company with huge ambitions.

In a small society with most of the population living in the capital it was relatively easy for Stefansson, as a member of a well-connected cultural elite, to cultivate government politicians. He had been at school with the Prime Minister David Oddsson, who as a neoliberal politician shared Stefansson's highly market-oriented vision of a technological and commercial cornucopia. In such a context there is no need for lobbyists. Oddsson publicly declared himself willing to sweep away any ethical constraints that might impede the advance of the new DNA technology – new times need new ethics. (From 2005 to 2009 he was chair of the Icelandic Central Bank, where his enthusiasm for light regulation and massive profits helped precipitate the devastating Icelandic crash).

Stefansson approached the Ministry of Health with the HSD proposal in November 1997. While the discussions with the ministry were proceeding, deCode realised its ambition of putting itself on the global biotechnology map. In February 1998, Hoffman LaRoche, then the fourth largest pharmaceutical company in the world, entered into a $200 million, five-year deal with deCode on positional cloning, that is, gene-hunting. The deal, as Roche's publicity emphasised, was solely directed towards the first of deCode's two business objectives and had nothing to do with the HSD. Roche offered up-front investment and payments for the achievement of 'milestones', although neither the exact nature of these milestones nor the size of the payment was made public. Nonetheless, the deal was big news in Iceland and in the global world of biotechnology. Oddsson publicly endorsed the deal, welcoming deCode as wealth creators, as bringing high-tech jobs to Iceland, and as offering the opportunity for young Icelanders working abroad to return home. Vigdis Finnbogadottir, who as President had herself spoken frequently of the need for high-tech jobs to bring young Icelanders home, shared Oddsson's enthusiasm for the new company.

In June 1998 deCode reported that it had started work on twenty-five common diseases. Two years later its website was recording that DNA analysis had started for twenty-eight diseases, ranging from arthritis to Alzheimer's and asthma. In 1998 the company claimed it had the capacity to map twelve common or complex diseases each year, and to have already fully mapped five. It emphasised its ability to take on new disease categories when requested by corporate partners, thanks to the technical ease offered by its highly automated systems able to run each sample through a battery of DNA analyses.

> Genotyping that is done on patients for a particular disease has value for all subsequent diseases that deCode genetics studies ... For example, deCode has a nuclear family that fits into extended family networks for diabetes, familial essential tremor and MS. Hence genotyping this one nuclear family provides information about all these other diseases.

Such open-ended gene fishing has long been a matter of concern for genethics as it bypasses the internationally accepted standard of requiring informed consent for genetic research carried out on human subjects. This norm is laid down in guidelines such as the Helsinki Declaration, the recommendations of the international Human Genome Organisation, and in European legislation. As we discuss in the next chapter, in the case of the Swedish biobank Umangenomics, getting the appropriate level of consent right was no easy matter. Stefansson, however, saw the information in the health and genetic records simply as an unexploited resource to be mined as a matter of moral and national duty. The legislation establishing the HSD echoed Stefansson in speaking of 'a duty to use the data'. Quite what Stefansson's admired colleague Milton Friedman would have made of invoking a moral and national duty in order to drive through a competitive enterprise is unclear.

The HSD not only divided Iceland's life scientists and clinicians, it also mobilised criticism from leading international geneticists. Endorsement came from the entrepreneurial culture, not science or medicine. Thus, in addition to the ringing praise given by *Red Herring* in the start-up days of deCode, and Hoffman LaRoche's early readiness to invest in the firm, Stefansson was welcomed by the Davos economic summit in early 2000. International scientific critics included the US geneticist Leroy Hood, himself no stranger to commercial funding, who expressed doubts about a monopoly licensee. The US breast-cancer geneticist Mary-Claire King, together with Stanford-based lawyer and ethicist Henry Greely, wrote to the Prime Minister of Iceland urging him to rethink. Canadian epidemiologist and World Health Organisation consultant A.B. Miller worried about the impact of such a monopoly licensee on the rest of academic genetic research. Strongly opposed to the deCode project as commodifying human genes, population geneticist Richard Lewontin urged that, subject to the views of Icelandic geneticists, there should be a boycott of Icelandic genetics. In California, the Icelandic physician Bogi Andersson and the systems biologist Bernhard Palsson raised concerns in both the general and the scientific press. Palsson was sufficiently angered by the prospect of the deCode monopoly that he established a rival biotech company, the Icelandic Genomics Corporation, in collaboration with a local entrepreneur, Trygve Lie, as a rival to deCode. Few UK geneticists other than Steve Jones in his *Telegraph* column spelt out their views quite so publicly, although concern about the Icelandic events was routinely expressed at meetings where the ethical implications of the new genetics were under discussion. Unquestionably the most hostile international criticism came from an editorial in *Nature Biotechnology* which, picking up on the case of a faltering and much criticised UK company, bluntly observed that British Biotech and deCode were the most conspicuous examples of 'how not to do biotechnology'.

Speed, the magic ingredient, clearly carried toxic side effects.

LEGISLATING FOR THE BIOBANK

The first time anyone other than deCode staff, key health ministry personnel, or senior members of the government heard about the proposed HSD was at a meeting on 23 March 1998, six months after deCode had faxed a draft of the bill to the government. On that day, fifteen biomedical experts, including senior geneticists, were invited at noon to a meeting to take place that afternoon at the ministry. There they were told about the forthcoming database bill. Dissatisfied, the experts demanded to see the written text. Two days later the text was made available – in confidence – to the fifteen experts. Comments were to be sent to the ministry before noon the next day, giving the experts less than twenty-four hours to produce a considered opinion. Such cursory treatment enraged many clinicians and biomedical researchers.

Eight days after the oral presentation to the expert group, Ingibjorg Palmadottir, Minister of Health, introduced the Health Sector Database Bill to the Althing (the Icelandic parliament). It was only then that, without any public consultation, the Icelandic public first learnt about the proposed database. The electronic database of the medical records of the entire population was to be funded and constructed by one licensed company, which in return was to have monopoly access, and the right to market the database for a period of up to twelve years for genetic discovery, diagnostics, drugs and health-care management. Universal consent was presumed, with no exceptions or provision for opt-out. The bill was criticised by the Social Democratic opposition as offering inadequate privacy; given the small population, identification would be a significant risk. Although the bill spoke only of 'the' company, and deCode was nowhere named, it was widely understood that this was a one-horse race.

The bill thus legislated for universal presumed consent. But while epidemiological research recognises that there are cases where presumed consent is necessary, this requires both an ethical and a research rationale. DeCode offered neither. Although the general public was by and large untroubled and saw the HSD in the same favourable light as the government, many of the relevant professional groupings – clinicians, nurses, biomedical researchers and human rights lawyers – found the bill ethically unacceptable. The core ethical, cultural and policy issues concerned consent, privacy and the impact on the health-care and research systems. Thus Tomas Zoega, Chair of the Icelandic Medical Association, pointed to the threat to the relationship of trust between doctor and patient posed by the database. If patients thought everything they said to the clinician would go into the database, they would stop talking openly. Medical records are primarily directed towards patient care; the secondary use of such records for research is increasingly recognised as risking breaches of privacy. Genetics and digitalisation accelerate the risk, but even counting prescriptions such as the retrovirals prescribed for HIV/AIDS can lead to the identification of patients with stigmatised health conditions. Psychiatrist Peter Hauksson perceptively argued that the supporters of the HSD were mostly people with no personal history of serious disease and did not understand the need for confidentiality in the doctor–patient relationship.

Furthermore, the easy assumption that there was no conflict of interest between non-profit and profit-making research glossed over an obvious problem. The failure to acknowledge and deal with this conflict of interest, not least around intellectual property rights, was a conspicuous weakness in the legislation, most clearly seen in the case of the already existing cancer database. Breast cancer was one of deCode's chosen diseases, yet Icelandic academic research in the field of genetics, using the meticulous registers of the national Cancer Society, had been crucial to the identification and characterisation of BRCA2 (a gene predisposing to breast cancer). What damage would the presence of a well-funded commercial biotech laboratory do to the capacity of highly successful non-profit Icelandic research teams to continue to deliver world-class genetics? The intellectual property debate was extended to the medical records themselves, with the government claiming that the state owned them. The opposition Social Democrats argued that the patients owned the records. The clinicians saw it as their legal responsibility as the guardians of the records to maintain confidentiality. This fight over the ownership and stewardship of the records led many clinicians to oppose them being transferred wholesale into the database. Forty-four GPs and 109 hospital specialists signed a statement refusing to submit clinical records unless requested by patients. This intellectual property debate, raging in a small Nordic island, occurred contemporaneously with the billion-dollar battle of the titans between the US NIH and the Sanger Centre on the one hand and Venter's Celera on the other (as we discussed in Chapter 1). In the age of intellectual property rights, genes have become big money.

The outrage which greeted the first bill from large sections of the clinical and biomedical research communities led to the formation of Mannvernd – the Association of Icelanders for Ethics and Responsibility in Science. Mannvernd was to play a central role in opposing commercial access to the medical records, setting up a website to mobilise international support, carrying translations of key Icelandic documents, media coverage and links to international scientific journals. A central figure within Mannvernd was Einar Arnason, professor of evolutionary biology and population genetics at the University of Iceland, who published extensively criticising the scientific assumptions underpinning the HSD and challenging deCode's claims of the genetic homogeneity of the Icelanders, precipitating a sharp debate in the international journals.

As well as the biomedical researchers and the doctors, there were human rights lawyers, a disillusioned major investor in deCode, and patients with chronic diseases. The patients suspected that the new managerial tools deCode sought to develop could and would be used by a neoliberal government to justify cuts in drug provision. Even deCode's offer to provide its own drugs at no charge was less than warmly received by such experienced patients. Along with many clinicians they were sceptical about the benefits on offer; would a new deCode drug for schizophrenia be more effective than the current drug and would the clinician have any choice over which to prescribe? Biomedical researchers were divided; some, recognising the cost, time and sheer difficulty of translation of research into drugs, were similarly sceptical as to whether there would be either genuine medical benefit to patients or economic benefit to Iceland.

The revised bill became law in December 1998 as the Heath Sector Database Act. But although Mannvernd remained unsatisfied, it had secured three substantial concessions. First, while presumed consent remained, the possibility of 'opting out' was inserted, though to do so required considerable effort. All households had been sent a pamphlet explaining the HSD and telling them of their rights to opt out, but the pamphlet did not include a copy of the opt-out form. The effort of opting out was left to the citizen. Opting out has predictably very different outcomes than informed consent, the former maximising the numbers participating, the latter significantly reducing them.

The second concession strengthened the privacy and confidentiality of the data. Crucially, personal identifiers were now to be encrypted; the licensee would not have a copy of the key; and the information would have a unidirectional flow. In this way, individual patient data could not come back or be attributed to the individual patient. In order to get the bill through, the government had sacrificed re-access. The Data Commission was given the responsibility for separating personal identifiers from the medical information, storing the cryptographic keys, and making sure that rare or isolated cases would not be featured in database analyses. In this way the legislation brought Icelandic practice into line with the European Union's Data Protection Directive (necessary for Iceland's participation in the European Free Trade Area). According to a deCode researcher cited by the anthropologist Gisli Palsson, resolving the privacy problems in this way created serious technical problems for deCode.[2] There would, it seemed, be no way of linking individuals' subsequent medical history to the DNA data. However, Arnason points out that because of the small size of the population and the genealogies it would still be possible to work back from the data and thus identify individuals. He quotes an interview in *New Scientist* in 2000 in which Stefansson himself says that it would be possible for deCode to identify a family with a disease of interest and then approach them for formal permission to cross-reference their names with the HSD.[3] (Arnaldur Indridason's internationally successful murder mystery, *Jar City*, turns on a deCode researcher making just such an identification. His mother's rapist was responsible for him being the carrier of a lethal genetic disease which leads to the death of his own small daughter and his murder of the rapist.[4])

The third concession reflected anxieties about the impact on the research system. In addition to giving monopoly control of the HSD to the successful tender, the bill also provided for the purchase of access to the data by biomedical researchers from outside the licensee's company. If these researchers worked for the Icelandic health service and thus contributed to the database, they would be charged at a preferential rate. However, this access was still conditional on the researcher's interests not being in conflict with the commercial interests of the company. That the licensee was to be both judge and jury on this matter of commercial interest did little to allay the fears of the non-profit sector researchers as this gave the licensee unreasonable powers to control access and thus academic research.

Although the legislation offered adult citizens the right to opt out it failed to say whether, once entered, they could withdraw their data, either on their own or their children's behalf. Like informed consent, the right to withdraw data is a standard element in ethical guidelines and frequently obligatory. This was to become a sore point. When the Council of Europe's Steering Committee on Bioethics asked the Icelandic government whether data could be withdrawn, the reply was that this was 'subject to negotiation'. However, this claim has to be set against the record of the debate in the Althing where the minister and government supporters had explicitly stated that data once entered could not be withdrawn. Further, the data on dead people (even those who in life may have opted out) were to be entered automatically and their families would have no say. Children under eighteen were given no say; their parents were to decide for them.

Again a significant deviation from ethical guidelines, and from what is in some countries a legal requirement.

The final legislation also failed to meet the ethical challenges raised by Mannvernd and the Icelandic Medical Association (IMA) over confidentiality and privacy. The IMA consulted Professor Ross Anderson, a Cambridge computer security expert who also advises the British Medical Association. Anderson's report was highly critical: stripping personal identifiers was not sufficient. In a country with 275,000 people, only a few pieces of data would be needed to reveal identity. He also bluntly drew attention to what he saw as deCode's 'lack of competence at computer security'. While Stefansson denounced Anderson as a 'hired gun', the IMA accepted his opinion and took the problem to first the Nordic then the World Medical Association, which also supported them. Europe's Data Protection Commissioners were of the opinion that the legislation might violate several European treaties, not least the European Convention on Human Rights.

The independent National Bioethics Committee, elected from the health professions and established as part of the patients' rights legislation, was opposed to the HSD, and had begun working on an alternative ethical framework, but it was suddenly replaced in August 1999 by the Director of Public Health. He directly appointed a new committee composed of civil servant professionals and a nurse as a patients' representative. The Director argued that the inability of the previous committee to make its mind up meant that change was necessary. However, it did not escape public comment that the new committee would be more malleable than the old.

So far as many clinicians and human rights lawyers were concerned, the previous year's legislation on patients' rights was now at risk from the database legislation. As we said, the Social Democrats argued, like Mannvernd, that morally patients should own their records, while the government took the view that the state owned them. With their huge majority the government easily passed the legislation but they underestimated the clout of the clinicians, as one of the most prestigious and best-organised professions. The doctors did not concede defeat. As professionals committed to protecting their patients' confidentiality they were not going to just hand over patient records which they held in stewardship. In his account of the struggle, Gisli Palsson fails to recognise the very different understanding of the records on the part of the government and of the clinicians. He, like the government, regarded the records as assets and describes the clinicians' opposition as the 'biopolitics of the dispossessed' – a conceptualisation ill-fitting a struggle between two powerful institutions over the ownership/stewardship of bioinformation. As the conflict worked itself through, it was Palsson's dispossessed doctors who won.

OPTING OUT

When interviewed in 1999, while it was still possible for citizens to opt out, Icelanders uninvolved in biomedical politics said that they had never seen the government's HSD pamphlet and expressed surprise on being shown it. Many politely speculated whether some other member of the household had thrown the pamphlet away, along with the junk mail. However the coverage by the widely read newspaper *Morgunbladid*, television and radio made ignorance of the issues, and not least the 'opt out', almost impossible. Palsson's respondents, interviewed rather later, told how they were willing to have their details entered into the database but that their doctors had refused to do so. Mannvernd campaigned for 20,000 opt-outs, and by November 2000 could announce that no fewer than 19,437 had opted out – at 7 per cent of the Icelandic population, large enough to potentially affect the validity of the HSD.

How the database was viewed tended to vary with gender, age and health status. Most people with a history of good health found it very hard to connect to the debate, retreating to a position of benign good intent towards less fortunate others. They felt they had nothing to hide, and no cause for health concern in the future, so for them there was no problem in participating. Others were less sure; they saw genetic information as potentially damaging to employment prospects, not least if the data were sold to employers. One was concerned about the implications for getting insurance, as Stefansson had said on the radio that deCode was interested in selling data to the insurance industry (a huge and largely successful fight against the industry took place in the US on precisely this issue). Others were critical, but not sufficiently so as to make them feel that they needed to take action. Thus one young woman in her late twenties, a flight attendant, said 'Well I don't want to know if I am going to die of a heart attack when I'm forty and I don't want anyone else to know either.' When I asked what she was going to do about opting out she replied that she 'almost certainly wasn't going to bother, it seemed all a bit unreal'. Other young women took a similar stance. The teenagers were uninterested; two sons of Mannvernd activist parents simply cracked up with laughter when asked whether they had ever discussed it among themselves or at school. They saw the HSD as totally irrelevant to them, although they were good-naturedly supportive of their parents' efforts.

The problems of opting out by members of vulnerable groups were raised in a discussion of the HSD as part of a course for professional carers for learning disabled people, some of whom would be deemed legally incompetent to make such decisions. In the discussion it became clear there was no policy to protect the rights of the vulnerable against the ethical problems posed by the database. In addition to their concerns about themselves and their children, the group, made up entirely of women, expressed their professional anger that the people they were trying to empower had been so disregarded. Their anger was matched by a will to restore the rights of the vulnerable while there was still time.

In November 2003 the Supreme Court ruled that including the data of the deceased in this way was unconstitutional. This ruling, on top of the 20,000 opt-outs, effectively destroyed any chance of the HSD succeeding in anything like its original form. Curiously many natural and social scientists, such as Herbert Gottweis (who also confuses sequencing and mapping), continued in their discussions of Iceland as if the HSD existed in any meaningful way after 2003, disregarding the significance of the ruling.[5] Such commentators have conflated the HSD with deCode's second biobank, constructed in collaboration with the clinicians, with informed consent and shared access to the clinical records. This was highly successful; deCode recruited some 140,000 Icelanders into a consensually built biobank doing well-regarded gene-wide association studies (see the next chapter).

THE SOCIO-ETHICAL CONCERNS OF WOMEN

What has been missing from most academic discussions of the HSD is the concept of gender, yet there is much evidence to suggest that women have the primary responsibility for their families' health. They are usually the primary carers of sick children and usually the parent who takes the time off work to do so. They also take main responsibility for their children's mental and physical well-being. These preoccupations were frequently spoken of by interviewees: would the deCode project really solve currently untreatable genetic diseases and, even more importantly, what might it mean for the future well-being of their own families? In the context of discussions with other women, where they felt they had time and space to explore the issues without being drowned out by the scientific or economic debates, women repeatedly raised their concerns for the psychological as well as the physical well-being of their family members. One woman with a history of breast cancer described how she first decided to opt herself and her young son out and then later struggled to persuade her father to join them. He was in favour of the HSD and erroneously saw breast cancer as a solely woman's disease. She explained to him that men could have breast cancer too, and reminded him that he had had a bypass. Supposing, she asked, this reflected a genetic predisposition to heart disease, did he really want to give his grandson the burden of the knowledge of this possible genetic risk as well as a possible cancer risk from her?

The claim made by the government and deCode that such feedback to an individual patient was impossible was beside the point; for her it was a matter of being cautious, of protecting her son, in a newly untrustworthy context. The commercialism of the HSD erased her sense of being cared for by clinicians committed to her and her family's well-being. Consent decisions that had belonged to her and her family members as patients were now going to be commercially presumed. She was not alone: women complained that their concerns of love and responsibility for their children were not being reflected in the media debates around deCode; instead, the gendered debate focused on issues of privacy versus progress and profit and, alongside the equally masculine preoccupations with fish quotas and sport, had occupied most of *Morgunbladid*'s pages all year. How could they begin to decide when there was no public discussion of the issues that troubled them?

Although ostensibly the database was to be encrypted with the information only flowing upwards, several of the women were unconvinced that the information would remain as purely statistical data and felt that, because of Iceland's small size, they could easily become identifiable. It was not the Cambridge computer security expert Ross Anderson's report that convinced them of this, but their own knowledge of how Iceland's telephone system had worked in the past (all calls went through the local switchboard operator who would listen in avidly). They also drew attention to Stefansson's promise during the general election campaign, where he appeared alongside government candidates, that database inputting would be put out to the local regional staff. What Stefansson saw as a smart move bringing employment to the regions, these women saw as creating local node points for the dangerous leaking of confidential material. Others felt confident in the guarantees given by the legislation that any leaks would be severely punished. Nonetheless, they repeatedly returned to the question of whether the knowledge of genetic information could harm their children – would it, for instance, make them fatalistic?

Women who had experienced domestic violence and sexual abuse raised some very different issues. One began to recall the past, when getting married meant showing your medical record to someone in authority – an indirect way of raising the spectre of Iceland's eugenic past. State eugenics, controlling women's reproductive capacities on the grounds of their fitness to mother, had not entirely disappeared from cultural memory. Without directly raising fears of a revival of the old state eugenics, they used this negative experience of medical records being used against women to raise their concerns about the implications of genetic information and surveillance for their children.

Another group of women passionately wanted to keep themselves and their children out of the database. They wanted as few people as possible to know their painful secrets. One woman, however, took a startlingly different position. She wanted her own and her children's records to be included, so that the database would have to record the sexual abuse both she and they had suffered. She wanted to end the social process whereby the victims of sexual abuse or violence hide themselves, or are hidden away in the name of privacy. If men fathered babies by sexually abusing their own daughters, as her father had, then it would be an act of justice for the DNA record to show just how many men had done this. There was sympathy and admiration, but no takers for her position. Some said that they had already decided to opt out even before they were invited to join the discussion. But all said that this was the first time they had had the chance to discuss frankly and confidentially what the database meant for them. And yet such discussion and reflection is the ethical heart of informed consent and indeed of the democratic accountability of biomedical science itself.

GENETIC NATIONALISM

By contrast with the women, most of the men interviewed, other than those associated with Mannvernd, shared the narrative of genetic nationalism. For them, the Icelandic genome was a unique resource, which could benefit Iceland's health and wealth – they saw their genes as the country's equivalent of Norway's oil. Regarding the exploitation of genetic knowledge of the nation's human bodies as the moral equivalent of the exploitation of regional mineral resources was something of an ethical innovation. This argument was positioned within a progressive nationalism still strong in Icelandic culture, yet in sharp contradiction with the growing possessive individualism of neoliberalism. Stefansson's own ethical justification for deCode's project was even more remarkable and contradictory given his role in the commodification of Iceland's bioinformation and its threat to privacy. Responding to a *Guardian* journalist in 2006, he attacked those who had opted out: 'It is insane to look at this as a threat to society. Privacy is not a primary right … To refuse cooperation is predatory behaviour. It is against the duty of Christian morality from which we have constituted Western civilization. It's your duty. It's the fundamental idea of socialism.'[6]

NASDAQ CRASHES

For the first two years deCode's scientific claims were made by press release. In April 1999 the *Wall Street Journal* reported a joint Roche–deCode claim to have found a polymorphism for arthritis. Publishing results this way is hostilely viewed by the international research community: it demands refereed papers. In this initial phase there was a tension between deCode's need to be seen producing patentable deliverables because of the pressure from investors and markets, and its need to win scientific recognition. While Roche was awash with money, and able to put up a huge investment, as part of the pharmaceutical industry it also expected deliverables. With Big Pharma, no deliverables, no money.

On 14 March 2000 came the bombshell of the joint Clinton–Blair statement questioning the legitimacy of patenting human genes that had been raised by Venter's commercial interruption of the hitherto public sector sequencing of the human genome. Despite the subsequent presidential backtracking, the Nasdaq Biotech index crashed by more than 40 per cent; by the end of the year it was down by 54 per cent. Biotech financial analysts began to speak more cautiously of deCode's prospects as it went to market in early June. 'So far,' said one, 'the concept behind the company is exciting but it is proving hard to bottle and sell.' Meanwhile the original venture capitalists – always high-risk short-term investors – having realised their profits, pulled out. The company was floated, and Icelanders were encouraged by Oddsson to invest at between $18 and $21 a share.

DeCode's shares peaked at $31 but then steadily declined to less than $1, with many Icelanders losing their life savings. With all the problems gathering around the HSD, culminating in the 2003 Supreme Court ruling, deCode had already begun to reconfigure. Now the emphasis was on conventional gene-hunting. Further collaborations with Roche were announced in 2004, and a stream of research papers reporting genetic associations for conditions as diverse as heart disease and schizophrenia appeared in high quality journals, but nonetheless as a commercial venture deCode continued to slump. It was not helped by the expensive and widely reported failure of a heart drug in late clinical trials.

In these increasingly tough market conditions, deCode again reconfigured its business, this time to cash in on the growing market for direct to consumer DNA testing and personalised genomics – deCodeme was established as a rival to 23andMe and other companies, offering to analyse DNA samples and provide direct to customers information on risk for forty-seven common diseases based on key marker SNPs. However, soon both 23andMe and deCode ran into difficulty. In 2010, 23andMe was rescued by Google, which became a major investor in the company. This did not help deCode; problems within the biotech sector were massively compounded by the global banking crisis and rumours of the company's impending bankruptcy were rife. Because of deCode's scientific success it was widely speculated that Wellcome would step in and buy the research wing of the company. Nonetheless in 2010 deCode filed for bankruptcy in Delaware. A company that had never made a profit ended up with total assets of $69.9 million against a debt of $313.9 million. By the end of the year it was delisted from Nasdaq.

BANKRUPTCY AND INTELLECTUAL PROPERTY

The bankruptcy of a genomics company raises issues of intellectual property. Who owns the information or even the samples in a DNA biobank and what happens to them in the case of bankruptcy? Could the biobank itself be disposed of like any other asset, and therefore without reference to the donors? Does it make a difference where the company is registered? As the deCode story further unfolded, the answers, at least for one jurisdiction, became clearer. Even as the bankruptcy was being prepared, Stefansson, with indomitable entrepreneurial energy, was planning a post-bankruptcy future. Sure enough, shortly after the bankruptcy had been filed, the Delaware court approved his rescue plan to sell a subsidiary company of deCode, Islensk Erfdagreining (EHF), to a newly formed group, Saga Investments. Investors included two of the venture capital firms that had invested in the original deCode project, as well as Illumina, a sequencing machine company in San Diego, California. Stefansson became executive chairman and president of research at EHF. Meanwhile, in an interview with the online journal *ScienceInsider*, Stefansson said that the parent company deCode genetics would be selling its drug discovery and development components. The new company EHF would focus on the search for disease genes and planned to begin whole genome sequencing of 2,500 DNA samples from its Icelandic database. He went on: 'It will not need to recontact these individuals for consent because their original consent agreements cover whole genome sequencing.'[7] To carry out this further DNA analysis, EHF had to have access to deCode's DNA biobank – both the tissue samples and the records. For the Delaware court at least, the answer to ownership had been provided; the biobank was treated as a transferable asset.

The rise, fall and re-emergence of deCode, albeit with clipped wings, over the past decade and a half has had its own specific trajectory. The mythic status of Iceland's unique genome and the intensity of Stefansson's vision and drive propelled its parabolic course. Kari Stefansson, like Craig Venter, is exemplary of the new hybrid – the life-scientist entrepreneur. Being hugely savvy in dealing with the media is integral to their success.

In deCode's wake, but for the most part lacking such charismatic leadership, a succession of national biobanks have followed. But while the US jurist David Winickoff's observation in 2006 that deCode's history helped produce the technological, political and normative terrain of all large-scale genomics initiatives today, and not just Iceland's,[8] the post-crash terrain remains unclear. The initial premise for the biobanks – the hope of personalised pharmacogenomics – has faded as the genetic associations that they are uncovering have turned out to be far more complex than the banks' promoters had envisaged. Just what constitutes a 'good population' for genetic studies has come under scrutiny. And the biggest economic crisis since the Depression of the 1930s has intervened. How long the recovery will take, and how much political unrest will be generated, are still unknowns. In the biobanks' preoccupation with what many social scientists speak of as the bioeconomy, the political economy of a globalised capitalism has been forgotten. In the next chapter we will chart the course of these other DNA biobanks through the uncertainties of genetics and the turbulence of the markets.

6

The Global Commodification of Bioinformation

Biobanks are not new. Collecting and storing human tissue has a long history. Scientific medicine's focus on such tissues, however, began not in the West but in the golden age of Islam, a period extending from the eighth to the fourteenth century, when scholars dissected the human body to study its internal mechanisms. Despite the mediaeval trade in saints' relics in the West, religious law forbade the practice of human dissection. It was only with the Renaissance that the Church's grip weakened, enabling both the arts and the sciences to dissect human bodies. In the centuries that followed, human dissections became for medical students a routine part of Western medical education, crucial for teaching the anatomy underpinning the surgeon's work.

Initially surgeons made their own collections of both normal and diseased and damaged human bones and organs. By the closing years of the eighteenth century these personal collections were largely located in the new hospital pathology museums. Skulls, skeletons, formaldehyde-preserved diseased and mis-shapen organs and foetuses were curated with the objective of both transmitting and advancing knowledge as part of the programme of modernising medicine. By the twentieth century wax-embedded tissue slices and frozen blood had been added to the collections. How many such human bio-specimens have been collected and stored is unquantifiable, but to give an example for Sweden, a small country with a population of 9 million, pathologists estimate that there are between 40 and 80 million paraffin blocks in store at any one time. All these collections could be called biobanks, but the term itself only became widely used with the advent of the new DNA biobanks as the millennium approached.

With the media's uncritical enthusiasm for the latest genetic explanation for this or that disease or behaviour, genetic determinism became the order of the day, marginalising the greater complexity of social, cultural and economic explanations of disease. The successes of the health promotion campaigns inspired by the World Health Organisation in the 1980s – which had reduced heart disease and premature death, dramatically so in some countries, notably Finland – were forgotten; gene-hunting was the order of the day. In this fin-de-siècle culture of genomania, the vision of what DNA biobanking could offer to science, medicine and wealth creation generated a global proliferation of DNA biobanking projects, from Iceland to the UK, from Northern Sweden to Quebec, Tonga, Norway, dozens in the US and many others in every researching country. Geneticists, venture capital, Big Pharma, and local and central government all wanted in on the dream of personalised medicine and new blockbuster drugs.

Early in 1999, while deCode was beginning to construct the HSD in Iceland, a proposal for a DNA biobank embracing the entire and highly heterogeneous population of Britain was published in the *British Medical Bulletin*, entitled 'Impact of Genomics on Healthcare'. The editors of this special edition included the Oxford professor of clinical medicine and biotech entrepreneur, John Bell, and George Poste, President for research and development at SmithKlineBeecham (SKB – later to be merged with Glaxo as GSK). The proposal was fleshed out in a paper in *Science*, by Poste together with Robin Fears, a biochemist and science policy consultant, setting out the case for 'building population genetics resources using the UK NHS'. In it, they made friendly reference to Prime Minister Tony Blair's claim that New Labour represented a 'Third Way' which sought to replace politics and political struggle with managerialism. Fears and Poste, noting the excessive speed which had stoked the blazing rows surrounding the Icelandic HSD, also took the advice of the World Health Organisation to proceed slowly, getting all stakeholders on board rather than rushing in and risking alienating swathes of doctors and many biomedical researchers.

Both deCode and Fears and Poste's projects were for DNA biobanks of two very different national populations, in the 1990s seen as the most promising route for population genomics. But there was an irony in that global Pharma, biotech, governments and venture capital all recognised that countries with a universal health-care system provided the ideal location for such projects. Since marketised systems such as that of the US exclude many from health insurance and from cradle to grave medical care, many also fall outside its recording system. In the US, even after the much vaunted Obama health-care reforms, the Census Bureau reported that 50 million Americans were uninsured in 2011, a million more than in the previous year. But other than the availability of comprehensive data, what constitutes a 'good' population remains numerically and genetically fluid. Thus, while deCode emphasised the smallness and homogeneity of Iceland and its highly educated population as 'good', Fears and Poste's proposal saw the larger, culturally and genetically diverse UK, with its educated population and half-century of medical records as an even better 'good' population – an issue to which we return towards the end of this chapter. They saw the absence of an internal market and its universality as making the NHS 'an unmatched but underleveraged vehicle that can put the UK in the vanguard of rational care based on molecular medicine'. They continue: 'The alternative bleak prospect is progressive rationing of increasingly constrained health care resources.'[1]

What this 'rational care' glosses over is the way in which, by identifying genetic 'risk factors', predictive medicine serves to expand the potential demand for drugs for people with no symptoms and who feel themselves to be healthy, making patients of those who may or may not have considered themselves as at risk. If the risk is high, as with single-gene disorders such as Familial Hypercholesterolaemia, and where there are clear and effective therapeutics (the statins), then living in this quasi-well quasi-patient state is one thing, but where the predictive power is low, as with many polygenic disorders, the therapeutics are likely to be unclear. By constructing a new kind of patient, the at-risk individual, personalised genomics creates a fresh market of unforeseeable size carrying with it an unpredictable but potentially vastly increased drug bill to be met by the NHS, and within privatised medicine largely by insurance.

There was a significant difference between Fears and Poste's proposal and Stefansson's ongoing project, which aimed to realise products and profits immediately, in that the SKB duo saw establishing a nationwide DNA biobank and health database as a pre-competitive phase in which all the major stakeholders – academic researchers, industry and government – would collaborate. Once the biobank was established as a research platform, then the data, protected by appropriate regulations, could be marketed. This difference as to when profit-taking should occur ran parallel with the very public debate over the HGP, with the demand for immediate patenting and profits by Craig Venter versus the no-patenting now and profits-later position of John Sulston and Francis Collins.

In the following decade and a half, beset by problems of ethics, property rights, confidentiality, public/private conflicts of interest and costs, some, like deCode, have risen and fallen in the space of a few years, while UK Biobank, sustained by infusions of public money, ultimately met its objective of establishing a research platform. But before turning to this last and most ambitious model involving the largest study population, we will follow the narratives of three other smallish biobanks, from Estonia, Sweden and Canada, all set up within months of each other. The first foundered by being unable to secure sufficient financing; the second through a failure to comprehend the problems of intellectual property rights; the third survived its stuttering start by recognising the necessity of being linked to a new international collaborative project.

ESTONIA

In 1999 the Estonian government seized on a proposal by Andres Metspalu, a professor of biotechnology at the University of Tartu, to create a DNA biobank which could, echoing the claims of deCode, 'do for the Estonian economy what Nokia had done for Finland'. Estonia, like Iceland, was deemed to have a 'good population' not this time because of any particular genetic homogeneity but because of its size and the availability of universal medical care and recording systems plus well-cared-for tissue banks. But while the Estonian political classes were enthusiastic and shared Metspalu's vision, a number of clinicians thought that patient care, particularly in the context of a weak economy with an underfinanced health-care system, should take priority over genomics. Their argument failed to win support and the project went ahead. Metspalu promised that the Estonian Gene Bank Project would learn from the Icelandic experience, would not sell data and would ensure that participating donors would have given their informed consent and would receive feedback in the form of a personal 'gene-card' which would help them reduce their predicted risk and enable their doctor to prescribe with greater precision. The recruitment of minors aged between seven and fifteen, and the fact that data once entered into the biobank could not be withdrawn, were less compatible with the ethical claims.

At the same time Metspalu set up a private company, Egeen International, registered and with its company headquarters in California. In return for exclusive commercial rights, Egeen provided most of the resources for the pilot phase of the sample collection, which aimed to include 10,000 donors. In an article in *Open Democracy*, the Estonian health researcher Tiina Tasmuth noted the scale of the financial inducement given to doctors to recruit donors for this pilot study.[2] They received 10 per cent extra pay for taking part, with 250 crowns for every hour spent with a donor, a considerable improvement over their basic salary of 42 crowns per hour. However, in the deteriorating climate for biotechnology of the early 2000s, the project fell into considerable financial difficulty and by 2004 had to be rescued through substantial underwriting by the Estonian government. It is now funded by the EU and located in Metspalu's department in the University of Tartu. Meanwhile Egeen has become a standard small drug-development company, with Metspalu as its Chief Scientific Officer. Essentially, the Californian company has bought the Estonian genome – bio-information moves smoothly within the global flow.

UMANGENOMICS: THE ETHICAL ALTERNATIVE?[3]

In 1999 both *Science* and *Nature* welcomed a new Swedish biotechnology company, UmanGenomics, as 'a model of how public tissue banks should interact with the biotech industry'.[4] The company was the brainchild of the University of Umeå and the County Council of Västerbotten, and was set up to commercialise the Medical Bank located at the University teaching hospital serving the entire county. The Medical Bank was already in existence, consisting of large numbers of blood samples drawn from 87,000 people, linked to clinical and questionnaire data derived from the WHO health promotion campaign begun in the 1980s. For the UmanGenomics planners, the Medical Bank was the realisation of the dream of the 'gold mine', a hitherto unexploited public sector biobank, just lying there waiting to be mined by a private genomics company. UmanGenomics seemed the ideal model. Yet just four years later, the company was more or less mothballed. At the end of 2002 there were sixteen full-time staff; by May 2003 there were but two.[5]

At the beginning, all seemed set fair. The plan for the company, to 'discover disease-related genes, explore their function and market this knowledge', was in line with the gene-hunting approach which had dominated molecular genetics so far. Professor Sune Rosell, recently retired from the Swedish drug company Astra, but with no background in genomics, was appointed as CEO. The plans were very deliberately 'ethical', compatible with both Swedish and European guidelines and regulation. The core requirement was for each and every new investigation to have fresh informed consent – an already obvious impossibility for genomics with its ever changing, ever faster technologies, able to handle huge quantities of data.

The UmanGenomics plan soon ran into difficulties: legal, marketing and scientific. The original public-private partnership plan provided for the University and the County, as the perceived owners of the bank, to own 51 per cent of the shares and for the rest to be privately held. Although the university was permitted – indeed encouraged – to develop industrial links, it was illegal for the County to engage in commercial activities, a problem glossed over by the neoliberal enthusiasm of the ruling Social Democrats. The marketing plan, which relied on securing contracts with major pharmaceutical companies to help develop the bank on a pre-competitive basis, failed. The science, too, had advanced beyond the scope of the initial thinking; UmanGenomics had to change its plan. Reorienting was costly and the company argued it needed some 200 million Swedish crowns (approximately £18 million) fresh public investment, now claiming that the 51 per cent public ownership was putting off potential investors. Ethics, which had been its selling point, had become a market handicap.

Despite a new contract being negotiated there was an ongoing fight over the legal status of the biobank's samples and patient data. This problem was built into the original commercialisation plans, for these were developed as if the Medical Bank was simply the result of the earlier health promotion activities of the County, as if the freezer chests full of blood samples, records of life-style, health status and such measures as lipid levels and blood pressure had come into existence without the input of biomedical researchers/curators and for no purpose. Yet it was acknowledged by all the actors in this story that the creation of the Medical Bank was the almost single-handed achievement of the director, nutritionist Professor Goran Hallmans. He had established and maintained the bank which he and his research colleagues and students had drawn on extensively as a basic research resource. But such biobank research had been unfashionable and did not easily attract funding until the mid-1990s when genomics came onto the agenda. In these difficult years Hallmans's research had been supported by the Swedish MRC and the Cancer Society, and had developed international collaborations as well as ingenious projects for unemployed people to find and train a labour force to take care of the bank. UmanGenomics planners had failed to consider Hallmans's possible intellectual property rights and simply assumed they had de jure ownership.

In the early 1990s, Hallmans had made an unsuccessful attempt to develop the biobank commercially. Later, together with a colleague, Professor Joakim Dilloner, he proposed an alternative social enterprise model in which ownership was to be vested in a foundation representing the people of Västerbotten. This was to be managed by the researchers, the County Authority and the University as well as by potential major funders such as the Wallenberg Foundation. All profit would be reinvested into biobank research. As both scientific researchers and social entrepreneurs they saw this as meeting the trust placed in them by the donors, benefiting research and honouring the Medical Bank's commitments to its existing funders. The University's and County's plans cut across this scheme.

This background helps explain how it was that a furious row broke out between Hallmans and the University as the lead owner of UmanGenomics, played out very publicly in the local newspapers, with many academics and even national politicians intervening. The case was followed closely by a multi-disciplinary research team led by Uppsala bioethicist Mats Hansson.[6]

At the heart of the conflict lay questions not of bioethics but of intellectual property rights. Unlike many countries (e.g., the UK), Swedish law allows for 'the teacher's exception', which gives academics, even though civil servants, ownership of the intellectual property they produce. However the UmanGenomics contract only recognised the County and the University as 'principals with rights of disposal' over the Medical Bank. Meanwhile legal uncertainties remained: did 'the teachers exception' apply to biobanks, and just who had the 'right of disposal' over this particular biobank? The conflict became bitter with little space for compromise. The University saw Hallmans as an impediment to its plans for both scientific progress and high-tech industrial development in Umeå. Hallmans, supported by a group of current and retired colleagues, was outraged that the contract between the County and the University set aside pre-existing research contracts, that UmanGenomics (like deCode) had monopoly commercial access, that the plan ignored the fact that donors had not given permission to have their samples and data exploited to make money for private profit, and that his intellectual property rights had also been ignored.

The University's Ethics Committee, still uncomfortable with the situation, continued to insist that the relation of the Medical Bank to the University's governing body was 'unclear and must be investigated'. They further argued that the University's proposal to appoint a steering committee of just four people to set the criteria for research based on the samples breached both the Committee's legal responsibilities and those laid down by the Swedish biobank legislation. Nonetheless, shortly after this, the University removed Hallmans from his post as director of the Medical Bank, though he was still insisting that he was the principal investigator and as such the authorised signatory to binding contracts. No small new biotech company could survive this, least of all in a stock market now reluctant to invest in anything other than very obviously robust biotechnology projects. Thus within four years UmanGenomics had become at best a 'virtual' biotech company.

So could it have been different? The difficulty of exploiting an existing public biobank via a start-up biotech company without the involvement of all the relevant stakeholders – not least the researchers who established the bank – strengthens the pragmatic conclusion of the commercial lawyers in Mats Hansson's team: 'because it is uncertain in Swedish law whether the biological research results initially belong to the scientist or to the university, our advice to an outside contracting party ... is to ensure that ownership is assigned to both'.[7]

These are legal, economic and research strategy issues; the UmanGenomics model, overly preoccupied with ethics and informed consent, was manifestly inadequate. Despite the 'ethical model' being hailed by *Nature* and *Science*, they missed the breaking point of the IP issue with its mix of legal and ethical problems. Arguably, given the idiosyncratic history of the Medical Bank's development, the Dilloner/Hallmans's model might have better secured the support of all the stakeholders from donors to researchers to the responsible authorities. Certainly less glitzy than the mistakenly admired 'ethical model', a social entrepreneurial model might have been better able to develop from the bottom up in this dynamic and multifaceted field.

CARTaGENE

As will already be evident, the DNA biobank enthusiasts saw an eye-catching name as an essential part of branding (the mix of lower and upper case letters is the sine qua non of commercial or would-be commercial genomics). Hence CARTaGENE. In 2002 a group of biomedical researchers based in Montreal and McGill universities and including the bioethicist and lawyer Bertha Knoppers, an international player in the genethics debates, proposed a biobank for Quebec. They argued that the Quebecois population, like that of Iceland, was genetically homogeneous, and therefore especially 'good' for DNA biobanking. Like Icelanders, the Quebecois could trace their genealogies back, if not a millennium, at least five generations (say 150 years max) since the arrival of their ancestors from France. The CARTaGENE project proposed to collect tissue samples, matched to health and socio-demographic measures and genealogies, from some 60,000 Quebecois volunteers, using appropriate consent procedures and at a cost of Canadian $38 million (at the time, the Canadian dollar, at around 0.6 of the US dollar, was at an all-time low as Canada had fallen behind its neighbour during the boom the US had enjoyed). Their initial bid for major funding from Genome Canada – a not-for-profit organisation established by the Canadian government with a $250 million budget to develop and implement a national strategy for supporting large-scale genomics research – was turned down. According to the McGill medical anthropologist Gilles Bibeau, while Genome Canada rejected the project as financially unrealistic, given the Icelandic furore there were also concerns about CARTaGENE's organisational structure and whether the project would be accepted by the public.[8] Nor, Bibeau argues – like Arnason in the deCode case – were the claims about the genetic homogeneity of the Quebecois valid. Unlike deCode, the conflict around the project never hit the global media, although it was intense and, like UmanGenomics, primarily local. Following the rejection, CARTaGENE's proponents successfully proposed a cut-down version which secured funding first from the Quebec state government and the University of Montreal, and, in due course, from Genome Canada. By 2010, having achieved its reduced objective of recruiting 20,000 participants aged between forty and sixty-nine (the same age range and completion time as UK Biobank), CARTaGENE announced that the research platform was now open for business.

Regardless of whether the Quebecois were indeed a genetically distinct population, even at 60,000 CARTaGENE would almost certainly have been too small a population for a stand-alone biobanking research platform. At 20,000 it certainly was. However, Knoppers was also a leading figure in the founding of the Population Project in Genomics (PPG), an international consortium working to harmonise data collection among the many small and large biobanks that were springing up. (All of which had begun by claiming the scientific legitimacy and social importance of their stand-alone projects.) Perhaps such small projects should more realistically be seen as a way of buying into global genomics and the vision of a biomedical and financial cornucopia.

UK BIOBANK

New Labour successfully fought the 1997 general election on a manifesto that contained no policy for science and technology. Indeed, Tony Blair disarmingly admitted during the campaigning that he didn't know how to send an email. Blair, it seemed, had little or no feeling for the potential of the new technosciences of biotechnology and informatics. Thus, despite his advocacy of the 'modernisation' of Britain, when appointing a shadow minister for science and technology in opposition, he had passed over biochemist MP Lynne Jones, already a member of the House of Commons Select Committee on Science and Technology, instead appointing Leeds MP John Battle, an arts graduate and anti-poverty campaigner. Not only did Battle face a sharp learning curve with a potentially vast and fast-changing brief, but he was also caught between his Catholicism (he systematically voted against abortion in the Commons) and his ambition to become science minister (where he would be responsible for funding human embryological research). Or was Battle simply an example of the new kind of MP who had begun to appear in the 1980s and dominates today – the professional politician with little or no expertise about anything in particular, except party politics, and with highly developed skills in political manoeuvring, campaigning, briefing and lobbying?

As part of its huge majority after the 1997 election, two new Labour MPs, both with senior experience of biomedical research and administration, entered Parliament: Ian Gibson, professor and dean of life sciences at the University of East Anglia, and biochemist Phyllis Starkey, a scientist administrator at the Biotechnology and Biological Sciences Research Council. Neither was brought into government. The clear message was that whatever part science and technology were going to play in shaping the future, the policy was not going to be formulated from within the Labour Party nor debated in the public arena.

Blair's steadily repeated mantra of modernisation called for the recognition that the British nation must change. His 'We will be a beacon to the world if the people unite behind our mission to modernise the country' echoed the tropes of Harold Wilson's 1964 speech 'forging socialism in the white heat of the scientific and technological revolution', but with a critical difference. With the abolition of Clause 4 of the Party's constitution, Blair had successfully overthrown Labour's long-held objective of taking industry into public ownership. With the residue of Labourist socialism eviscerated, there was now little ideological difference between Thatcherism and New Labour. (Making it even harder for his successor, the less than decisive Gordon Brown, to nationalise Northern Rock during the banking crisis of 2008).

Devoid of any research agenda of its own, New Labour was more than open to proposals being made by powerful figures in molecular medicine, finance and industry that their joint collaboration with the government would improve the nation's health and wealth in equal measure. As sociologist Michael Rustin observed in the journal *Soundings*, what was new about Blair's New Labour was that in the history of the Labour Party it was:

> the first time the power of capital and the markets which empower it is regarded as merely a fact of life, a reality to be accommodated to, and not a problem, not a force to be questioned and resisted. The abstractions of 'globalisation', 'individualisation', even 'informationalism', can be used to reify the real agents and interests which dominate the contemporary world.[9]

Just so, and Rustin's analysis fitted New Labour's research policy making like a Savile Row suit. Real agents and interests were dissolved in the language of getting all the stakeholders together under the 'big tent'. Elected experts – particularly Gibson, who was to become chair of the Commons Select Committee on Science and Technology, but who were seen as not fully signed up to New Labour – were less welcome than unelected enthusiasts.

Disbelieving in any structural antagonism between capital and labour – or, more modestly, between the different interests of employers and employees, or those of a globalised industry and the future of Britain – the Blair government was more than open to the grand imaginaries offered by genomics and informatics. Nowhere was this better seen than in Battle's replacement in short order by David Sainsbury, of the immensely wealthy Sainsbury's supermarket family, who had developed what he described as his twin passions of politics and science while a Cambridge undergraduate in the '60s. He had joined the Labour Party early on, but unhappy with its growing internal chaos had switched allegiance to the new Social Democratic Party. On the losing side in the internal SDP struggle to remain independent or merge with the Liberals, he switched again, becoming a supporter and major donor to New Labour. Having risen to CEO in the family supermarket chain in the 1990s, he had front-line experience of the GMO conflict in which biotechnophiles like himself were defeated by the Europe-wide popular revolt against 'Frankenfood'. In the aftermath, supermarkets and caterers (including the Sainsbury chain and the outsourced caterers provisioning Parliament and the Royal Society) found it necessary to ensure – and publicly declare – that their foods were GM free.

Sainsbury's twin passions were realised when Blair appointed him as Minister of State for Science and Innovation in July 1998. His path to success had not been hindered by his donation of a million pounds to the Labour Party in September 1997, which had been followed a month later by his elevation to the peerage. These events triggered a fierce row, with Mark Seddon, a member of Labour's National Executive Committee, denouncing Sainsbury as having bought the Party. Certainly he has been the largest single donor in its history. Meanwhile a depoliticised New Labour read Sainsbury's donations (amounting to £11 million over the years) and his refusal to accept a ministerial salary as simply part of his acknowledged generosity in donating to educational and research projects. Sainsbury was to become indispensable to New Labour, serving as minister for longer than anyone except Blair and Brown.*

The first the public learnt of the existence of a UK DNA biobank project was in the autumn of 1998, through the BBC programme *Newsnight*, broadcast when the deCode story hit the media. Wellcome, unwilling to let the Icelandic story get all the publicity, ensured that news of the much bigger UK Biobank found its way into the newsroom. The leak ran that it would include 500,000 adults – more than twice the size of the Icelandic bank – that it was being jointly planned by the Wellcome Trust and the MRC, and that in principle they had agreed a preliminary figure of £25 million, again rather more than twice the money Stefansson had secured from the venture capitalists backing deCode.

For those outside the powerful network assembling itself around the UK Biobank project – that is the elite scientists, the funders and Big Pharma – this was something of a surprise. In 1999, the Department of Trade and Industry, under Sainsbury's auspices, published a report on the economic potential and strategic importance of biotechnology in the UK, with the evocative title of *Genome Valley*. The report argued that following the IT revolution, biotech was next in line, and that the UK's strengths in basic research and pharmaceuticals made it strongly placed to benefit. But

* Sainsbury's enthusiasm for Blairism survived the fall of the Labour government in 2010; he is now a major financial sponsor of Progress, a Blairite think tank opposed to much of the programme of the Party's current leader, Ed Miliband. In the merry-go-round of capitalist globalisation, Sainsbury's supermarket chain is now a quarter owned by Qatar's sovereign wealth fund.

both R&D and venture capital are mobile, and Britain needed to be made an attractive place to invest through fiscal incentives. This required more public investment in universities and basic research, and, interestingly, moves to 'increase public confidence in science'. In line with New Labour's embrace of capital, twenty-five of the thirty-two members of the working groups involved in producing the report came from pharma and biotech companies. The remainder were predominantly from the ministries and Research Councils. But the first glimmerings of UK Biobank appeared even before *Genome Valley*, with the horizon-scanning 1995 *Foresight Report* which had already identified the genetics of common diseases as a major area for investment. It foresaw that the early ability to predict individual risk would give the UK biotech and pharma industries a leading edge, poised to exploit the anticipated 'strong customer demand' and the 'high export potential'.

BUILDING THE BIOBANK

If deCode was the hare among the DNA biobanks, UK Biobank has unquestionably been the tortoise, taking twelve years from the first audible leak in 1998 to the completion of data collection in 2010. But even if the hare was bankrupted, has the fabled power of the slow but steady tortoise really won out? Although the research platform has been built, the question remains open as to whether it can deliver scientifically and commercially. And is health care likely to be transformed as so many of the leading figures have predicted?

DeCode and the Estonian Gene Bank Project had been driven forward by charismatic scientists: Stefansson and Metsapalu. UK Biobank lacks any such figure, though the ambitions of the Wellcome Trust cannot be discounted. Instead the proposal flows through a mesh of committees, in the corporatist model in which the presence and close cooperation of all the main stakeholders is seen as ensuring the smooth growth of globalisation. From the beginning it was recognised, not least by the private sector, that the biobank could not be developed by them alone – any more than could the HGP. *Foresight* encouraged the state and charitable trusts to fund what was described as the pre-competitive phase of research. Like the HGP, UK Biobank was about building a research platform, not actually carrying out research. While all the stakeholders including the private sector get to help pick the scientific and technological 'winners', the initial risk entailed in platform-building is to be carried by the public sector; if the project does indeed become a 'winner', the gains are to be shared. And yet, even acknowledging the significance of the involvement of Wellcome Trust, already a powerful interventionist in the HGP, just how did Britain, under a New Labour government with no visible policy for science and technology, come to commit itself to such a major scientific project?

Within the mesh of committees from Foresight onwards, the names of key players appear again and again: Sir David Cooksey, Sir John Bell and Sir Christopher Evans. In an increasingly neoliberal Britain, such figures rarely represent just one interest; not for nothing does Helen Wallace, the Director of *Gene Watch*, call them the Biotechnology Barons.[10] (*Gene Watch*, a very visible presence in the story of UK Biobank, is an influential public-interest research group funded by the Rowntree Reform Trust, one of the very few non-charitable trusts in Britain which funds political campaigns to promote democratic reforms, civil liberties and social justice.) Among the pioneering UK biotechnology barons is engineer and entrepreneur Cooksey. Having established Advent, the pioneering British venture capital company in 1981 – almost the only non-American company investing in deCode – Cooksey was and remains a leading figure in both the public and private sectors, chairing major governmental, academic and industrial committees, including the National Audit Commission for twenty-four years before becoming a governor of Wellcome from 1995 to 1999. Cooksey headed the new British Venture Capital Association in 1995, and during the following decade chaired Diamond Light Source Ltd, the UK's synchrotron, in which Wellcome had a significant shareholding. He retired as Chairman of the Directors of the Bank of England in 2005, then chaired a Treasury Committee reviewing UK health research funding in 2006 to strengthen translation research.

Professor Sir John Bell is both a member of the Board of Biobank UK (a charitable company) and Chair of its Scientific Committee, as well as Chair of the Office of Strategic Co-ordination of Health Research. A founder of three biotech companies, he is also a member of the board of directors of Roche and its associated biotech company Genentech. Imperial College Professor Sir Christopher Evans founded biotech firms, chairs the venture capital firm Merlin, and worked closely with Tony Blair before and after the 1997 election – not unlike David Sainsbury. These biotech barons move seamlessly between advisory committees to the UK government and Europe, academic and industrial laboratories, national and international scientific and industrial groupings, and between the governing bodies of the major governmental and charitable funders.

This consensus among the baronry that UK Biobank was a near certain 'winner', and Wellcome flexing its formidable financial muscle, meant that the biobank planning committee was pushing at a half-open door. In 2001, the influential House of Lords Select Committee on Science and Technology gave its support, overcoming objections from the Department of Health (DoH), which at that point preferred localism for bioinformatics in the NHS, as this is directed towards improving health care at the local level. By contrast nation-centred bioinformatics is primarily directed towards research, and in particular UK biobank. Localism lost out and by 2002 the DoH supplemented Wellcome's and MRC's initial £25 million with a further twenty. The appointment of epidemiologist John Newton as CEO in 2003 did not work out. From the beginning his grasp on the project was less than convincing; when he went to brief the Parliamentary and Scientific Committee about the biobank he insisted that it was not about predicting individual risk but group risk, and that it would make not just pharmaceutical breakthroughs possible but also methods of providing social and emotional care to the sick. Newton was soon replaced by his Oxford colleague, Rory Collins, another epidemiologist, and UK Biobank began to pick up speed.

PLANNING UK BIOBANK

Putting flesh onto the *Foresight* outline began with the discussions between the MRC and the Wellcome Trust. The joint committee, chaired by the London School of Hygiene epidemiologist Professor Tom Meade, had already decided that Poste's nationwide project was premature and had reduced the proposed population a hundredfold. UK Biobank's advocates spoke of its unique potential to disentangle the contribution of 'nature' and 'nurture' (or genes and the 'environment') in explaining the major killers – the common diseases. A sample of 500,000, more than double that of deCode's, would, they considered, be adequate. The figure never changed, despite pointed criticism from statistical geneticists and others.

The initial scientific protocol thus envisaged inviting a population of half a million aged forty-five to sixty-nine to participate, to be recruited through participating GP practices. The participants would donate blood and urine samples, have some basic biodata recorded, and complete a health questionnaire ranging from their weight at birth to a record of their diet over the previous seven days. The protocol specifies the size of the cohort required to provide enough cases to study particular diseases and conditions as they developed over the subsequent years; thus it would take thirty-one years to collect 5,000 cases of bladder cancer, fifteen for hip fracture, thirteen for Alzheimer's disease and only four years for diabetes. The half-million upper-middle-aged participants would thus provide the 'good population' for building the biobank.

Meade's committee saw that winning and maintaining public trust was vital, hence there needed to be an independent ethics and governance committee. The health policy consultant William Lowrance was appointed to chair this second planning group. Lowrance's interest in the secondary use of biomedical data and privacy was seen as directly relevant to issues confronting UK Biobank. ('Secondary use' means data gathered for patient care being reused for research.) Despite security being flagged up as a concern, at no point was a data security specialist appointed either to the scientific or the governance committee.

The committee also began public consultation, commissioning a market research company to conduct focus groups and opinion surveys. Meade and his colleagues also spoke at public meetings. This attempt to ensure public trust in the biobank was probably helped by the establishment of the Human Genetics Commission (HGC) in 1999. The Commission had held regular open meetings where the complex ethical and social aspects of the fast-changing field of genetics were discussed. But while the HGC included distinguished academic social scientists with relevant expertise, UK Biobank did not. As a result neither its planning committee nor their market researchers seemed aware of the Economic and Social Research Council's programme for 'The Public Understanding of Science' carried out a decade before. The case studies carried out by the sociologists in that programme provided the evidence which led to the abandonment of the model of 'public understanding' and its replacement by 'public engagement' – that is, to a move away from the one-way-street model in which the job of the scientists was to explain the science, and that of the public to listen and learn, to a two-way-street model of dialogue. The sociologists' case studies did not automatically privilege the accounts of science above the lay people they interviewed; what they were concerned with was how lay people understand a specific science-related issue, whether radioactivity in Cumbria, a genetic predisposition to heart disease, or the risks of living near a chemical factory.[11] This approach made it possible to explore the distinctive contribution of experiential knowledge to the science-related issues of the case studies. By contrast, the market researchers, very much in the grip of the one-way-street model, reported one of its focus groups as having shown 'a raft of unhelpful negative attitudes to genetic research' and 'having views based sometimes on misinformation and mistaken assumptions'. Rather than giving voice to those they studied, the market researchers at times felt entirely able to judge what their subjects said according to its conformity to science.

Although the public was brought into the consultative process early on, not only were their views sometimes discounted but the big questions were never put on the table. Did UK Biobank offer value for money? Would it improve the health of the UK's multicultural citizens? What were the opportunity costs and risks? Some of these concerns had been raised in the focus groups; for example, would the time spent by the GPs working with UK Biobank reduce their time for patient care? Would insurance and tobacco companies be permitted access to the completed biobank? Such specific concerns were met only with reassurances that as a charitable company UK Biobank would only be allowed 'to act in the public good'.

UK Biobank's management structure is built around three bodies: UK Biobank Board, the Scientific Committee (now the Steering Group to the CEO), and the Ethics and Governance Council, this last seen as guarding the public good. At the time of writing, the Board includes representatives of the funders, MRC and Wellcome, is chaired by the ex-chair of NICE, and includes five biomedical researchers, an accountant, two civil servants from the Scottish and the England and Wales departments of health, and a lawyer who had earlier been involved in planning the Ethics and Governance Council. But, like the original planning committee, it includes no social scientists. The Steering Group consists solely of biomedical researchers. The Ethics and Governance Council is chaired by a lawyer and includes four bioethicists, two social scientists, the retired CEO of a patient society, an epidemiologist, a geriatrician, and a retired engineer with an interest in citizen participation. While there is overlapping membership between the Board and the Steering Group, there is no overlap of the Ethics and Governance Council with either. In the context of the trust issue there has been little or no attempt to innovate either in the structures or their membership – there is no equivalent, for example, of the Community Advice Council established by one US DNA biobank, where lay people can discuss the issues in an independent forum.

THE SCIENTIFIC PROTOCOL: MANAGING THE CRITICS

The initial scientific protocol was sent to twelve anonymous international referees, chosen by the planning committee, who were asked to comment on its international competitiveness, timeliness, value for money and study design. UK Biobank resisted the request from the House of Commons Select Committee on Science and Technology to see the reports, which were released only when *GeneWatch* put in a Freedom of Information request. Despite being chosen by the biobank's proposers themselves, and therefore likely to be broadly supportive of the project, the referees expressed a number of serious concerns. They agreed that the project was timely and that the UK was a good place to locate it (though two referees suggested that Scandinavian countries might be better because of their more comprehensive health records). In terms of value for money, there was some surprise that it could be done so cheaply, with the referees questioning the assumption that half a million tissue samples and biosocial data could be collected, curated and stored at a cost of no more than £100 per case – compared with the usual rule of thumb figure of £1,000.

However, the study design itself was subject to tougher criticism. Two referees questioned the whole concept of a prospective study, arguing that the classical approaches of family study and case-control would be cheaper and more effective. Three raised concerns over the proposed forty-five to sixty-nine age group, arguing that this would miss many early onset diseases and, importantly, pre-menopausal women (the final protocol acknowledged this criticism and lowered the catchment age to forty). There were also major concerns about the sampling and population size. The protocol envisaged that 40–50 per cent of those contacted through their GPs would respond; one referee thought this unrealistic and that 25 per cent was a more likely level. (In the event it was much lower, as we discuss below.) Furthermore, some referees thought that the recruitment process would mean that volunteers were likely to be among the healthier of the population, and that ethnic minorities would be under-represented. Above all, there was doubt over whether 500,000 would be big enough – only 50,000 of either sex in each age cohort, as one referee pointed out, not enough for adequate statistical power. The referees were particularly concerned that no statistical or epidemiological geneticists were associated with the project – even going so far as to suggest the names of those in the

UK who should have been consulted. Despite being described as a project to disentangle 'nature' from 'nurture', the 'environmental' or phenotypic data that would be collected would miss crucial factors such as earlier exposure to pollutants – and just how many older people could be expected to recall their birth weight accurately? Faced with such criticisms, it would seem unlikely that, in a period where only a fraction of even alpha-rated projects were being funded, a review committee not already committed to the project would have agreed to support it.

Independently, UK geneticist Sir Alec Jeffreys (creator of DNA fingerprinting techniques) argued that the cost of the proper collection and curating of the samples would run into billions of pounds and that the danger of false positives was a serious hazard to interpreting the data. The publication of the referees' reports and a critical paper by statistical geneticists David Clayton and Paul McKeigue[12] prompted an aggrieved joint response from Michael Dexter, Director of Wellcome, MRC's CEO George Radda and Tom Meade.[13] *GeneWatch*, however, drew on these criticisms in its submission to the House of Commons Science and Technology Committee to argue that UK Biobank's aims were controversial, its science questionable, and that there was a lack of legal safeguards for volunteers on the potential misuse of their genetic information. The Committee, chaired by biologist MP Ian Gibson, went a long way to accepting the critique, arguing that the biobank had been approved without obtaining a proper scientific mandate. As the MRC had failed to allocate grants to alpha-rated research proposals, funding the biobank had severe opportunity costs for biomedical research. A report published in *The Lancet* in 2003 claimed that among the biomedical researchers it had canvassed, 'the only point everyone seems to agree on is that sufficient debate about the project has not taken place'.[14] Gibson raised the issue at a brief Commons adjournment debate but was mollified by the promise of a full debate – which in the event never took place.

THE WINNING TORTOISE

It was not until 2007 that recruitment began, following some minor revisions in response to the reviewers and pilot studies involving several thousand participants, now relabelled 'volunteers'. The initial protocol documents had assumed a 40–50 per cent response rate but the pilot studies showed that only 10 per cent of those invited to participate did so. No one in the planning group seems to have questioned this, even though the refusal of nine out of ten people is extraordinarily high. What made the refusal rate particularly noticeable was that subjects had been invited through their GP practice, a procedure which typically ensures a higher rate than a cold-call random sample. A study of the refusers could have cast light on how the research should be modified to achieve the projected 40–50 per cent response rate. Was it a practical matter, that they simply had no time? Were the refusers dissatisfied at being offered no feedback on their own health? Were they concerned about their lack of control over the potential users of the data? Had the concurrent scandals around the abuse of the crime DNA databank affected their trust in DNA biobanking? The DNA databank, which the police were determined to hold on to, included genetic information taken not just from those found guilty of crime but also from victims, those interviewed during crime investigations, children under ten and people who had been tried and acquitted. The human rights issue was taken to the European court, the crime databank declared in breach, and, with ill grace, the UK government made improvements. Studies of refusers from other biobank projects suggest that non-responders are most likely to be disproportionately poorer, younger or from ethnic minorities (or any combination thereof), distrustful of the beneficent intentions of the research, or have reduced mobility. A factor common to such passive or active refusers is that most are likely to be living in worse conditions and poorer environments than the volunteers.

To compensate for the low response rate, invitations were sent from thirty-five collection centres across the UK to no fewer than 5 million men and women in the age group living within ten miles of a centre. Volunteers were asked to complete a touch-pad questionnaire – giving information on their smoking and drinking habits, current job, postcode and car use (this as a proxy for environmental pressures), plus some basic biodata: height, weight, hip measurement, blood pressure, lung capacity and handgrip recorded – and also to provide blood and urine samples. It was these environmental measures, such as the postcode and life-style data, which enabled Rory Collins to argue in response to Alec Jeffreys that UK Biobank was an epidemiological rather than a purely genetic project, albeit that the referees had earlier identified the protocol's concept of 'the environment' as both inadequate and imprecise. Although Jeffreys accepted Collins's argument, it remains the case that the social information collected in the questionnaire is relatively 'soft', thus the only really 'hard' data are the biomarkers, and above all the one measure that is (barring mutation) stable over a person's lifetime – DNA. This, plus the unrepresentative 'sample', once again pushes explanation towards genetics.

The information collected is strictly limited by the budget, and the entire process must last no longer than ninety minutes (the pilot studies took 100 minutes, adding substantially to the costs). Samples are stored for up to twenty years and, if funds become available, it is hoped that the health and disease histories of a 25,000 subset of the cohort will be followed up every two to three years. Academic researchers and pharmaceutical companies wishing to make use of the research platform will put forward a proposal to UK Biobank, to be approved both by the Steering Group and by the Ethics and Governance Council.

THE SPINE

For biobanks to work, they must have access to an effective electronic medical record system, acceptable to the clinicians who will use it and to the patients whose data it records. By the end of the 1990s GPs had, with government support, established electronic medical records to facilitate best practice, intended to be shared by hospital and General Practice care alike, hence with a primarily local focus.

When DNA biobanks were under discussion by the House of Lords Select Committee on Science and Technology, they visited Poste at SKB and were won over by the research case for converting these localised records into a highly centralised IT system initially estimated to cost some £6 billion. This, it was claimed, would open up entirely new possibilities for biobanking research – at the expense of the clinical commitment to localism. The way was cleared for this data centralisation in the relatively unnoticed Section 60 of the 2001 Health and Social Care Act, which introduced assumed consent for patients' details being entered into the electronic record. This secondary use for research purposes of data originally gathered for patient care raises a significant ethical problem, as the appointment of the health policy specialist Lowrance to plan the Ethics and Governance Council had indicated.

Despite the caveats from the clinicians, by 2002 the government had adopted the idea and commissioned the construction of the NHS 'Spine', as it was named, the biggest public non-military IT project in the world. Essentially the Spine was a version of the Icelandic HSD, covering the entire population, and like the original HSD, there was to be no opt-out. After a bitter fight in which a patient who insisted on opting out was formally denied access to medical care (her GP took no notice of this), opting out without penalty was permitted. However, constructing the Spine turned out to be a great deal more difficult and costly than had been thought – IT companies were hired and fired, and twelve years and £12 billion of public money later, it had still not materialised. The seamless flow of clinical data has yet to arrive in the clinic.

From its beginning the concept of the Spine came under heavy criticism as its centralising model had been driven through without adequate consultation with health-care professionals. Furthermore, while the government and its IT advisers claimed that this highly confidential data would be secure, its committees had not included a biomedical data security specialist, drawing trenchant criticisms from the Cambridge computer security expert Professor Ross Anderson, the British Medical Association's adviser who had damned the Icelandic HSD. Anderson now encouraged a nationwide 'Big Opt-Out' campaign. Patients were urged to inform their GPs that they did not wish their data to be uploaded into the NHS's centralised system, a request many clinicians – as in Iceland – were happy to go along with. The anxiety about privacy was compounded by identity theft becoming a matter of widespread concern – in 2006 even Blair's medical records had been hacked into. The much publicised losses or theft of government departments' and NHS computer discs containing confidential personal files and information did nothing to inspire confidence. And as the Big Opt-Out campaign pointed out, some 400,000 NHS staff would have smart cards for accessing the data, offering both the likelihood of significant card loss and the possibility of leaks for cash.

But while civil libertarians were concerned about the loss of privacy in a culture where every card purchase gives yet more data to companies without consent, many had given up, seeing the digital society as having pretty much destroyed personal privacy. Quite separately, leading academic IT experts doubted the Spine's technical feasibility. The 1997 Labour government had been convinced by Cooksey, Evans and others that the country's future prosperity lay in fostering cutting-edge technologies, above all IT and biotech, through large injections of public cash. Brown was at least as enthusiastic as Blair. Yet as grandiose IT project after IT project overran in cost and time or proved technically infeasible (the debacle over computerising passports being only one prominent example among many), the likelihood of the Spine materialising receded until it was closed down by the incoming Coalition government in 2010, only to be resurrected in a new form as the Summary Care Record.

Helen Wallace's meticulous documentation of UK Biobank indicates that for some of its proponents, the 500,000 population sample had always been seen as a dry run for Fears and Poste's full 59 million. And in line with this prediction, in early December 2011, Prime Minister David Cameron announced that the NHS records would be opened up and made available to private companies as what he claimed would be an unparalleled resource for genomic research and drug development. There would be no opt-out, though data would be pseudonymised. This still leaves open a possible legal challenge, since, as patients' organisations were swift to point out, if postcodes, dates of birth and demographic details are all available, anonymity is far from secure. Prescriptions alone can be strong pointers – the retrovirals pointing to HIV/AIDS and the psychotropics to mental health conditions. Many would prefer such matters to be kept private. There was no indication of whether private companies would be required to pay for the privilege of mining the data. However, the government rhetoric matched that of Stefansson's view of the HSD. It is the 'communal duty' of patients to voluntarily contribute their data, for the 'good of the nation', and presumably in gratitude for the continued existence of an NHS 'safe in the hands' of the Coalition government.

There was also to be a further sweetener to public sector research in the form of a £186 million boost for translational biotech research and development. According to Allegra Stratton, former political analyst for the *Guardian*, the government's move came as alarm bells began to ring early in 2011, when several Big Pharma companies – including Pfizer, GSK and Novartis – announced the closure of major research facilities (Novartis's site at Horsham is in the Chancellor George Osborne's constituency) with the loss of many skilled scientific and technical jobs (of which more in later chapters).[15] Cameron's advisers convened a seminar on genomics, addressed by the business consultancy McKinsey, the biotech baron John Bell, and Wellcome's CEO Mark Walport. Biotechnology is the UK's third largest contributor to economic growth, with 4,000 companies employing 160,000 and with a combined turnover of £50 billion. Big Pharma's primary commitment is to their shareholders, so when a research field is judged unlikely to succeed, companies close it down. The record of successive UK governments over the past century in terminating failing large-scale technoscientific projects is conspicuously weak. Will this £186 million be good money after bad? That is, just what health or wealth benefits, in the light of the past decade and a half's experience, might UK Biobank supported by a nationwide electronic medical record be expected to generate?

THE GOOD POPULATION

In the years during which population-based DNA biobanks have been in construction, most attention has focused on the management, science and ethics of the banks. Far less has been said about whether they could, even under the best of circumstances, deliver the health-care benefits that their advocates have so strongly claimed for them. Rather, it has been taken for granted that as genetic knowledge increases these benefits will become evident as the data piles up. Genes will be identified and the public-private partnerships between the biobanks and the pharmaceutical industry will enable the development of personalised medicine. Yet there has been a steady critical chorus, not just from those sceptical of such overly gene-oriented approaches to public health, but from within the genetics community itself. Initially, the concern was over what constituted a 'good population' enabling genes associated with common or complex diseases to be identified. Back in the 1990s, at the birth of deCode, Stefansson argued that the Icelandic population was ideal for the proposed studies because it was genetically homogeneous. Whatever the legitimacy of that claim – and as we pointed out in the last chapter, it was strongly disputed – the broader question is whether such homogeneity does indeed make a 'good population' for genetic studies. Might it not mean that any results found would apply to that particular population and not to a more diverse one? This was the methodological (as distinct from ethical) critique directed at Cavalli-Sforza's Human Genetic Diversity Project and which led to the choice of mixed urban populations for the International HapMap project we discussed in Chapter 1. The merits of heterogeneity in the population was one of the main arguments advanced by Fears and Poste for a biobank project involving the whole population of Britain, an argument they lost with the final 500,000.

On what basis, though, can one decide how big the study ought to be and how, if it is not to involve the entire national population, should its participants be recruited? One obvious criterion is that the sample must be big enough to include a sufficient number of people with the disease or condition of interest to make genetic analysis possible. A second is that it must be representative of the population from which it is drawn. As both the planners of UK Biobank and the project's referees recognised, this presents a serious challenge. In a population with many small ethnic minorities of diverse bio-geographic ancestry, does the design ensure that they are adequately represented? Ninety-six per cent of the participants in all the biobank studies to date are of European descent, and some have now turned the critique of Cavalli-Sforza's HGDP on its head by arguing that this both deprives much of the world's population of genetic justice, while also missing an opportunity to identify rare gene variants that might be present in non-European populations that either increase or diminish the risk of particular diseases.[16]

Disease patterns vary between classes, and – since the groundbreaking epidemiological studies of the health of the starving Rotterdam population in the final months of the Second World War – it has been known that the poor nutritional status of a woman at the time she conceives can seriously influence the health of her baby. Not only do regional patterns of morbidity and mortality differ across the UK (men in London's Kensington can expect to live fourteen years longer than men in parts of Glasgow), but so do gene frequencies – even, for instance, between people living in north and south Wales. As the initial referees of the UK project recognised with their emphasis on the need for statistical and epidemiological geneticists to be involved, the 90 per cent failure rate among those invited to opt in to UK Biobank ensures that it cannot provide a good population in the sense that Fears and Poste had originally anticipated.

DNA biobanks begin with the simplifying assumption that genes play a causative part in the aetiology of common diseases which may appear to run in families but where no single gene seems to be crucially involved. To take a non-clinical example of such complexity: in any given environment, height is rather strongly heritable, but there is no single 'height' gene.** There are more than ninety

** The fact that there has been a steady increase in average height in countries like the UK and Japan over the past century is an indication that heritability does not imply destiny; change the environment, and the heritability changes too.

different single nucleotide polymorphisms (SNPs) in different regions of the genome each marking a gene variant that has a small effect on a person's adult height. In the case of the common diseases, how large a population must be sampled to discover such variants depends on two factors – how frequent the disease is in the population and how many genes might be involved in its causation. For the specialist biobanks created to study a specific condition, such as the international Autism Consortium, the first problem is resolved as samples are collected from those with the diagnosis, and then compared with those from matched controls (case-control).[17] But for the population-based prospective DNA biobanks, those diagnosed with the disease must be extracted from the entire cohort and then matched with controls (an approach called 'nested case-control'). The gamble that the population biobankers took was that for common diseases like coronary heart disease (CHD), the number of significant genes would indeed be small – perhaps no more than five. On this basis, and with the known incidence of CHD in the population, enough affected subjects to enable the different genes to be teased out could be found within a biobank numbering in the hundreds of thousands. In so far as a genetic as opposed to economic and political rationale can be found for UK Biobank's half-million sample, this, it would seem, is it.

But as we discussed in Chapter 2, and as the gene sequencers have been forcibly reminded in the decade following the HGP, there are many reasons why heritability is not simple. The underlying assumption of the DNA biobanks is of the primacy of the genome, if only because DNA profiles and SNPs are hard data, whereas measuring the 'the environment' produces soft data. Epigenetic processes are superimposed upon the readout of the genome, epistasis (gene–gene interaction) occurs, but is poorly understood. And despite the lip service paid to 'the environment' by those modelling genetic effects, their formulae have a manifestly false built-in assumption that the effects of genes and the environment are almost entirely additive and can be teased apart by appropriate statistical methods.

Such doubts might have been dismissed by the optimists as mere carping, the sort of nay-saying that had presaged the start of the

HGP. Over the last five years, the results from the gene-wide-association studies (GWAS) that the biobanks have made possible have begun to be reported in the scientific literature. By 2011, over a thousand such studies had been reported. Each major paper – for schizophrenia, for arthritis, for ischaemic heart disease – has been greeted with high priority publication in the journals and press releases announcing that the gene or genes 'for' the condition have been found. But sadly for the hopes of the biobankers, it has not been a case of three, four or five major genes but of scores or even hundreds. A consortium of over a hundred authors reported that ninety-five genetic loci have been found related to blood lipid levels,[18] possibly hundreds of genes might be implicated in coronary heart disease,[19] and around a hundred in schizophrenia.[20] Each of these genes may contribute a fraction of 1 per cent to the probability of contracting the disease. In all, more than 1,200 loci associated with more than 165 common human diseases have been identified. With so many genes involved, and with the complex biochemical pathways that they engage, the pharmacogenomic prospect of personalised medicine – of drugs tailored to your particular genomic profile – recedes.

One of the few biobank successes has been in helping to explain why some people have adverse reactions to particular drugs, making it possible to screen before prescribing.[21] In a very few cases a DNA scan will give a positive indication of what to prescribe. Around half those with hepatitis C have a SNP close to a gene, *IL28B*, which means that the disease is treatable by a year's course of a complex drug regime – though the regime is so tough that not all patients are able to take it.

The response of the biobank enthusiasts however, has been to argue the case for increasing the population size of the banks, by combining the many national ventures now underway – for example, to create a Europe-wide biobank, resourced through the EU-funded Biobanking and Biomolecular Resources Research Infrastructure (BBMRI), slated to be up and running in 2012. But as one of the most trenchant geneticist critics of population DNA biobanks, David Goldstein, has pointed out, identifying even more genes with smaller and smaller effect sizes, many of which may turn out to be spurious associations, while it will continue to yield publishable scientific data, is not likely to be able to make any contribution to health – or even provide potential targets for the pharmaceutical industry.[22] Furthermore, even when the identified risk factors are combined on the dubious assumption that they are additive, they don't account for all the observed heritability. In some despair, geneticists have begun to speak of 'hidden heritability', of 'black holes' present in the genome – not genes that are causally related to the diseases, but other unknown factors undetectable by current methods.[23] However, as Eric Lander and his colleagues have pointed out, this may be a case not of hidden heritability but of phantom heritability. That is, what appears to be heritable is in actuality the consequence of interactions between genes and gene pathways during development – epistasis. Modelling Crohn's disease, where seventy-one risk-associated gene loci have been identified, accounting for only 21 per cent of the heritability, they show that interactions between just three biochemical pathways could account for the condition – but to identify them would require sample sizes in the order of 500,000 cases – a far cry from UK Biobank's assumptions.[24]

According to a press release issued jointly by Wellcome, the MRC and the Department of Health, in April 2003 the then CEO of UK Biobank John Newton told the Parliamentary and Scientific Committee that UK Biobank would 'allow the risk of disease to be predicted in populations', would 'specify meaningful subgroups of illness and improve the specificity, effectiveness of all kinds of care, not only drugs, but also social and emotional', and would 'show which proteins or biochemicals are present before disease develops and therefore may be causative'. He cited the Framingham study in the US showing that people (here Newton was euphemistic as the Framingham study was just of men) with elevated cholesterol are more likely to develop heart disease, which led to the development of cholesterol-reducing drugs like the statins. 'If the UK Biobank generated just one finding of this significance it would have been worthwhile in public health terms.'

Newton's choice of Familial Hypercholesterolaemia (FH),

caused by a genetic mutation affecting one in 500 people, makes an instructive case, though not one that supports his argument. Long known as a genetically transmitted disorder, a family history of premature death from heart attacks being the indicator, there was for many years no effective treatment, and clinicians often did not tell patients of their suspicions. With the availability of medication, suspicions could be confirmed by blood tests showing dramatically raised cholesterol levels. Today FH is routinely and reasonably successfully treated by two classes of drugs, the statins and ezetimide, together with life-style measures.

It was work by Michael Brown and Joseph Goldstein, comparing tissue cultures of liver cells from control and FH homozygous patients, that made the scientific breakthrough. By providing the target for drug treatment this opened the door to effective treatment and to them receiving a Nobel Prize in 1985. Although Brown and Goldstein were researching a genetic disease, their explanation of the mechanism came via biochemistry not genetics. Since then, researchers have discovered some 700 mutations for this single-gene disorder, which has so far produced interesting science, but not a new blockbuster more effective than the statins. Thus to the treating clinician the presence or absence of DNA information makes little difference – it would increase costs but not improve care.

Eight years after Newton issued UK Biobank's prospectus, Lander and his colleagues conclude that the belief that such genetic studies will lead to personalised risk prediction and medicine is now untenable:

> The proportion of phenotypic variance explained by a variant in the human population is a notoriously poor predictor of the importance of the gene for biology or medicine. A classic example is the gene encoding for HMGCoA reductase, which explains only a tiny fraction of the variation in cholesterol levels but is a powerful target for cholesterol lowering drugs [the statins]. Ultimately, the most important goal for biomedical research is not explaining heritability – that is, predicting personalised patient risk – but understanding pathways underlying disease and using that knowledge to develop strategies for therapy and prevention.[25]

It is not then surprising that in a critical article in *Science*, geneticist James Evans, health psychologist Theresa Marteau and colleagues point to the messy relationship between the scientific assessment of risk and its translation into clinical practice, and conclude that 'it is time to deflate the genomic bubble'.[25]

Others, remembering Thomas Kuhn's account of how, in the history of physics, there were regular attempts to save unworkable theories by bolting on increasingly implausible special factors, might conclude more simply that the genetic paradigm that underlies the DNA biobanks has failed.

7

The Growing Pains of Regenerative Medicine

One of the first acts of the incoming Obama presidency in 2009 was to overturn the 2001 ruling of his predecessor that no federal funds should be used for human embryonic stem cell (hESC) research other than on existing cell lines. Within three days of Obama's inauguration, US approval was given for the world's first clinical trial based on the use of such cells – a therapy for spinal cord injury developed by the Californian biotech company Geron, already owner of the Roslin Research Station in Scotland where Dolly the sheep had been cloned.[1] Geron's share price briefly doubled, while the President's decision was greeted by the London *Guardian* with a euphoric editorial:

> the potential of stem cell research is almost biblical in its scale. The capacity for these cells to transform into whatever the body needs to regenerate itself could, in the lifetime of the next generation, make the blind see, the crippled walk, and the deaf hear. It could cure cystic fibrosis and arrest muscular dystrophy. Yet it also raises one of the most difficult dilemmas in medical research: ensuring that something which could offer huge rewards to the nation's health is not overlooked because it does not offer huge rewards to the national economy ... [This] puts the US back in the race ... and, crunch permitting, scientists anticipate a surge of investment. Although the gap between this basic research and a therapeutic process is huge, if it works, the scope for patenting the manipulated cells will make its backers rich.[2]

The editorial reprises the then Health Minister Yvette Cooper's claim nine years earlier that stem cells from cloned human embryos 'could prove the Holy Grail in finding treatments for cancer, Parkinson's disease, diabetes, osteoporosis, spinal cord injuries, Alzheimer's disease, leukaemia and multiple sclerosis, transforming the lives of hundreds of thousands of people'.[3] Such hyperbole is characteristic of the stem cell saga which since the 1980s has been one of hopes raised and dashed, dramatic medical claims subsequently retracted, downright fraud, unethical research, untested therapies, stem cell medical tourism and regulatory regimes varying from the robust to the non-existent. By contrast with the Bush ban, most European countries had adopted the thinking of the Warnock Report, namely that the embryo did not have the moral status of a living child but that its potential for life did require special moral consideration, and must be regulated for accordingly.

The *Guardian* leader writer went on to suggest that the alternative approach of using the patient's own adult stem cells (called autologous), rather than embryonic ones, to regenerate damaged tissue is less likely to generate wealth 'since it is (rightly) impossible to patent an individual's stem cells'. The leader writer had evidently forgotten the game-changing 1980 Diamond v. Chakrabarty ruling, discussed in Chapter 1, that 'anything under the sun that is made by man' can be patented – at least within the US. In 2005 the FDA classified autologous stem cells as drugs, and broad stem cell patents, valid until 2015, are held by the Wisconsin Alumni Research Foundation, giving them the legal right to exclude anyone else in the US from making, using, selling or importing hESC. Although these rights only apply to the US, stem cells developed elsewhere will nonetheless fall foul of the Wisconsin patents if used in the US.

An EU Directive of 1998 banned 'the use of human embryos for industrial or commercial purposes' and the patenting of inventions deemed to be immoral or unethical. This compromise permitted the work to proceed, but set limits to commodification. In the same year, the German Patent Office granted a patent to Oliver Brustle, a neuroscientist and stem cell researcher, for a technique enabling stem cells to be turned into precursors for nerve cells. Greenpeace challenged it, and after more than a decade of legal challenges and appeals, in October 2011 the European Court of Justice ruled that no process or product that involved the destruction of hESC could be patented. The European stem cell research community was thrown into disarray. A group of twelve leading European stem cell researchers, headed by Austin Smith of the Wellcome Centre for Stem Cell Research in Cambridge, expressed their 'profound concern' at the recommendation on the grounds that to bring research into clinical use needed the involvement of the biotech industry, which would require patent protection.[4] Three of the twelve authors declared competing financial interests. Other stem cell researchers suggested that as it was only the patents and not the research that had been banned, work could and probably would continue with companies resorting to industrial secrecy rather than patenting. Given that universities, let alone companies, routinely and successfully lure high-status researchers with higher pay and more attractive conditions, and that their knowledge goes with them, this suggestion looks to be less than realistic.

Over the three decades since the possibility had been floated that hESC could be used to repair damage resulting from injury and disease, there have been intense ethical and religious debates. Pro-lifers, whether Catholic or fundamentalist evangelicals, insist that an embryo has the same moral status as a living child. To kill an embryo whether by abortion or as part of the research process is to commit murder. Many secular feminists also opposed the research, but for an entirely different reason; they are concerned with the reproductive freedom of women, not the moral status of the embryo. For the international feminist network Hands Off Our Ovaries, the issue was that embryos come from, and can only come from, women's bodies, which were now to be mined for research material. The alliance of the IVF pioneers Robert Edwards, an embryologist, and Patrick Steptoe, a gynaecologist – the latter therefore the gatekeeper to women's embryos – is replicated in the hESC teams. The ethical problem is whether the firewall between the clinical care of the woman as patient and the same woman as a potent provider of the egg is sufficient. Does a woman longing for a baby feel a sense of obligation to her gynaecologist? Does her informed consent have that 'voluntariness' at the heart of the Nuremberg Code?

For some years, as the result of an effective alliance between the religious, Green, feminist and secular opposition, and with the support of the leading philosopher Jürgen Habermas,[5] Germany maintained a position of total opposition to stem cell research as incompatible with human dignity. This central moral concept is enshrined in the German federal constitution, largely constructed by the victorious allies in the aftermath of 1945, and which echoes the United Nations 1948 Charter of Human Rights by placing human dignity and human rights at its heart. Attempts to bypass this prohibition were until recently blocked. Thus when the Prime Minister of the Land (i.e. state) of North Rhine-Westphalia visited Israel in 2001 in order to establish hESC research collaboration between Bonn and Haifa, where the regulatory regime is much softer, his initiative commanded little popular support and was firmly rejected by the parliament. Over the following years Germany – pushed by the researchers, encouraged by those in favour of wealth creation, and increasingly legitimised by a new generation of bioethicists less engaged by the concept of human dignity – cautiously negotiated its way from total opposition to support for hESC. These negotiations were analysed by sociologist Barbara Prainsack and her colleagues as a process in which 'Ethical reasoning was transformed into instrumental reasoning about the value of research. Questions that had seemed everyone's property became, in other words, issues reserved for a technical elite.'[6] Ultimately the new criterion was that of 'good science'.

The sociologists miss the bathetic repetition of history. The confident insistence that the embryologists were merely doing 'good science', just like any other biomedical researcher elsewhere, reprises the defence of the Nazi doctors on trial at Nuremberg. What does not seem to have been pointed out is that Nuremberg, and Helsinki following it, spelt out that without overwhelmingly compelling reasons no research should be conducted on humans that had not first been tested with animals. And for many of the conditions the hESC researchers have been offering to target, such as motor neuron disease – one of the first hESC research projects to be licensed in the UK – there are already good animal models that could have been exploited before testing on human embryos. hESC researchers spoke openly of how in their discussions with politicians they had emphasised that, given the strength of the animal rights movement, working on humans was less politically sensitive.* As a consequence, researchers working with animal stem cells complained that they found it harder to obtain funds than those working with hESC.

By 2006 only a small number of countries, including Austria, Cyprus, Italy, Ireland, Lithuania, Norway and Poland, continued to ban all hESC research. Most other countries either permitted in vitro research with embryos with no mention of stem cells, limited creation of cell lines to 'surplus' embryos, or imposed other restrictions. Obama's overturning of the Bush ban added the US to the list – although the permission is still under legal challenge in a number of lower courts. Belgium, Japan, Singapore, South Korea, Sweden and the UK, however, took a new and significant step by explicitly permitting the creation of embryos for research purposes.[7] The robustly regulated but highly permissive research regime in Britain has, as we have emphasised, been shaped by successive UK governments' priority that research should be directed towards wealth creation. Estimates that the market potential of hESC would reach $8.5 billion by 2020 with a substantial slice going to UK plc have been a powerful incentive. This relaxed view was endorsed by libertarian bioethicists but criticised by many European countries as permitting too much too fast.

Substantial government investment in hESC research raised the prospect of the migration of leading US stem cell researchers to the UK. However, although some did migrate to Britain during the Bush years, in practice the US ruling was less than binding, because it applied only to funds distributed by the Federal government. Individual States, foundations and industry were not restricted. In 2004, with Republican Governor Schwarzenegger's enthusiastic

---

* The political sensitivity of successive UK governments to the animal rights lobby was exemplified by the refusal to grant Colin Blakemore, a prominent defender of animal experimentation and CEO of the Medical Research Council, the customary knighthood that goes with the appointment.

support, California voted for Proposition 71, which launched a 3 billion dollar research bonanza in defiance of the Bush ban. Several other US States followed. The vote was followed by the creation of the Californian Institute for Regenerative Medicine, chaired by real-estate developer Robert Klein, who had led the successful lobby for the proposition. Though the Institute has been lavishly funded, it has had a somewhat bumpy administrative track record and currently a somewhat uncertain future. Research and researchers are footloose and migratory, following the money through the globalised enterprise that science has become. In the aftermath of the Obama decision, US stem cell researchers in Britain, such as Stephen Minger, correctly predicted that as UK financial support was so limited, they would consider returning to the US (he himself left his university position for Amersham, a British biotech company), and if not to California, then to Singapore or China, where by 2011 some thirty research teams, with more than 300 PhD researchers, were working on the cells.

Thus by the end of the first decade of the twenty-first century – as the wilder claims of genomics for the transformation of medical care had not been realised – the hope/hype which a decade previously had enveloped the Human Genome Project had been transferred to hESC.

WHY hESC?

The hESC promise has been built on the bringing together of three related lines of biomedical research and development stretching back over a century: study of the processes of embryonic development; the development of tissue culture techniques for first preserving and later cloning cells; and the possibilities of treating wounds or diseased tissue by transplantation. The two latter techniques were pioneered on animals and transferred to the clinic, while embryology was pursued for its first half-century as basic science. Research took place within a passionate philosophical debate between vitalists and mechanists. Vitalists argued that development involved a specific life process – an élan vital not reducible to physical and chemical mechanisms. However, by the 1920s the mechanists had won out. Embryonic development was, they concluded, a physical and chemical process.[8] So much so that by the 1990s, Lewis Wolpert, a leading developmental biologist, argued that the embryo was 'computable', a simple reading out of the instructions inscribed within its DNA program.[9] If only that were the case, the HGP would have yielded everything its advocates had claimed for it.

Reviewing the discoveries of this seemingly basic science shows how it became a forerunner to today's stem cell industry. The question the embryologists asked was simple to pose, but hard to approach experimentally. If a lizard loses its tail, it can grow a new one (a property called plasticity), but humans cannot grow new limbs. Efforts to find ways of repairing such damaged tissues have a long history. Why are mammals not like amphibians or reptiles? In mammals, just as in amphibians, embryonic cells have the potential to develop into any adult cell; but this capacity, partly preserved in amphibians, is lost in adult mammals. True, some human body tissues, such as skin or liver, can self-heal if the damage is not too great, with newly formed cells replacing the damaged or dying ones. But for most, such as heart and brain, the damage, once done, is permanent. Hence the immediate cause of many diseases and disorders lies in the death or injury of specific cells. Heart attacks damage the muscle cells in the heart. Parkinson's disease results from the death of a group of nerve cells in a region deep in the brain, the substantia nigra. Destruction of nerves in the spinal cord results in degrees of paralysis varying depending on the level of the cord at which the damage occurs. Dead cells are replaced with scar tissue.

To understand why this loss of plasticity occurs, late nineteenth- and early twentieth-century biologists began studying the processes of development, initially in free-living eggs from the sea urchin, later in frogs and toads, where the transformations from fertilised egg through tadpole to fully developed adult could easily be observed under the microscope and experimentally manipulated by teasing out and removing individual cells with fine tweezers at varying stages of development. The general rules formulated by this research were that, shortly after the moment of conception (the fusion of egg and sperm), the fertilised egg begins to divide, so that (in humans) after about five days it has become a ball of some hundred embryonic cells called a blastocyst, the precursor to the embryo. In the very early stages of cell division, an individual cell can be dissected out from the rest, without affecting the development of the remaining cells into a fully functional adult – a result cited by the vitalists as in their favour: the organism apparently had the power to overcome the injury. By contrast, removing cells later in development results in lasting damage to the developing foetus – a result that pleased the mechanists as it meant that the wounded embryo could not reorganise itself. It took many years of patient experimentation to resolve these seemingly incompatible findings. The answer, as so often in biology, is that 'it all depends' (among other factors on the stage the ablation is made and the organism chosen for the experiments). A caution biologists share with those who study human social systems.

The consensus conclusion was, and remains, that in its early stages any individual cell in the blastocyst can give rise to any adult cell type – it is totipotent. These are stem cells. However, as they divide, they gradually become more restricted in their developmental range, first pluripotent, and then finally committed to their fate as nerve, muscle or whatever. (While 'commit' and 'fate' are used as technical terms by developmental biologists, their wider resonances still reflect the vitalist/mechanist debate.) Thus over the nine months from conception to birth, the initially undifferentiated cells of the blastocyst are transformed into the many specialised cell types (over 200) that the human body contains – liver, muscle, brain...

These adult cells haven't lost the DNA required for cell division or differentiation, which is still present in the nucleus of every cell of the body (red blood cells are the exception as they don't have nuclei). However, the cellular control mechanisms that regulate which genes are active at any time have switched off the currently unwanted genes by some process in the cytoplasm – the enzyme-packed region of the cell that surrounds the nucleus. It is thus the state of the cytoplasm that determines whether a cell is totipotent, pluripotent or fully differentiated. In cancers some of these genes are switched on again, enabling the cells to multiply dangerously. But otherwise only a small number of stem cells remain – notably in bone marrow, from which blood cells are constantly being born, and in a part of the brain called the olfactory bulb. What it is in the cytoplasm of most adult mammalian cells that ensures such regulation, and whether it could be manipulated to restore embryonic totipotency, is even now something of a mystery. If techniques could be discovered to 'reprogram' the cytoplasm so as to recover the adult cells' lost pluri- or totipotency, then the clinical need for hESC would disappear, and along with it much of the ethical debate. Indeed, one consequence of the Bush decree was to stimulate intense research in the US on just this possibility. In the event, and ahead of the US researchers, it was a Japanese group led by Kyoto pharmacologist Shinya Yamanaka that made the breakthrough, first in mice in 2006, and a year later in humans – to which we return later. By making adult human skin cells once more pluripotent his technique was hailed as defusing the ethical debate around embryonic stem cells. Yamanaka called his cells induced pluripotent stem cells (iPSC).[10]

By the mid-twentieth century, new techniques had been developed making it possible to keep cells alive more or less indefinitely in tissue cultures. Embryonic cells, spread onto nutrient jelly in a Petri dish, and kept comfortable in a warm oxygenated broth, can grow, divide and differentiate. Suitable tricks can prevent the differentiation, so that the cells can divide indefinitely but remain in a totipotent or pluripotent state – so-called immortalised cells: another of those culturally resonant metaphors with which the field is full. Such 'cell lines' from lab animals became a staple laboratory tool, joined from the 1950s by similar lines derived from human cancer cells, notably HeLa, whose origins, embedded in good science and dubious ethics, we discussed in Chapter 3.

REPAIRING DAMAGE BY TRANSPLANTING CELLS

The final strand in the tangled web that led to such enthusiasm over the potential of hESC derives from the potential use of such cells to repair tissue damage. The idea that it might be possible to replace damaged or diseased tissues with healthy substitutes transplanted from other people or animals has a long history. Technically the easiest to administer is clearly blood, and there are ancient myths about blood transfusions restoring vigour to a dying man. Within Western medical history, however, the first recorded experimental attempts date from the seventeenth century, notably in experiments made on dogs by Christopher Wren (better known as the architect of St Paul's cathedral) under the auspices of the then infant Royal Society. The results were not encouraging; the animals became sick and died. Effective blood transfusion for humans had to wait until the twentieth century, with better understanding of blood groups and immune responses, while transplantation of tissues and organs was impossible before the arrival of techniques to lower immune reactions and prevent rejection.

However, there was a much older tradition, deriving from Indian medicine, of using the body's own skin tissue to repair noses. A flap of skin from the upper arm was partially detached, and grafted to the person's nose. After some fourteen days the flap could be cut from the arm and shaped to reconstruct the nose – the reshaping took from three to five months. The tradition was brought to Italy as early as the sixteenth century, but it wasn't until the exigencies of the First World War and especially during the Second that skin graft techniques became of central importance in treating burn victims. The pioneer in the field, Harold Gillies, recognised that to rehabilitate people whose faces had been disfigured required making it possible for them to look recognisably like their old selves. (The recent technical success of face transplants is existentially enigmatic. Given that the procedure has only recently been undertaken, even with human flexibility the outcome for people's sense of self-identity must be uncertain.)

Up until the 1950s successful grafting required a person's own tissue – so-called autologous grafting. Many of the problems of rejection of adult transplants and grafts from 'foreign' tissue were resolved following the work of Peter Medawar in London in the 1950s and '60s. The first successful kidney transplant was carried out in the US in 1953, and the first – controversial – heart transplants in the late 1960s by Christiaan Barnard in South Africa. But even earlier, Frank MacFarlane Burnet in Australia had shown that, because embryonic tissue could not distinguish between its own and 'foreign' cells, it was immunologically more 'tolerant' and therefore potential transplant material. Burnet and Medawar shared a Nobel Prize in 1960 for having opened the door to transplant technology.

Obvious targets for repair strategies are the brain and spinal cord. For many years it was believed that nerve cells in the central nervous system cannot regrow or be replaced (it is now recognised that this is not entirely true), so that damage or destruction of brain or nerve cells was considered to be irreversible. For example, unlike in the periphery, severed nerves in the spinal cord do not regenerate, and depending on how high up the cord the injury occurs, a person may be paraplegic or quadriplegic. One reason for this failure is that in adults the protective cells that surround the nerves in the spine block their regrowth. This isn't the case in embryos however, so for several decades researchers have been hunting, with little success, for techniques that would 'reprogram' the adult protective cells, or, alternatively, to replace them with embryonic tissue that would permit regrowth. But the appeal of trying to use hESCs to help the spinal cord regenerate was obvious – not least to one of its best-known campaigners, the paralysed actor, ex-Superman Christopher Reeve.

PARKINSON'S DISEASE AND THE MADRAZO AFFAIR

Another prime candidate for attempting to restore function by cell transplantation is Parkinson's disease, caused by the death of nerve cells in the substantia nigra region of the brain. These cells, when alive, secrete the neurotransmitter dopamine, and drug treatments for the disease have involved giving L-dopa, a chemical precursor to dopamine, to replace the missing transmitter. Parkinson symptoms can be mimicked in rats by destroying the substantia nigra, as a result of which the animals show tremors and movement disorders similar to those experienced by sufferers from Parkinson's. As far back as the 1970s, researchers had been exploring the effects of embryonic or foetal tissue grafts to restore function in such animal models of Parkinson's disease. In 1983, a team of researchers led by Anders Bjorklund in Lund in Sweden claimed that tissue taken from foetal rat brains and injected into the brain of aged rats restored motor coordination.[11]

These and related experiments generated intense controversy among researchers in the field. How did the grafts work? It was clear that the injected cells survived, but did they make new functional connections with other cells, so replacing the dead or damaged host tissue, or did they simply function as a sort of biological mini-pump, secreting the missing dopamine? Some even questioned whether the cells were necessary at all; Don Stein, in the US, reported that it was enough to simply make a hole in the head, insert the cells, and then remove them, to stimulate some recovery of function.[12] Meanwhile, in parallel to the animal work, Olof Backlund, a surgeon at the Karolinska hospital in Stockholm, was experimenting with autologous grafts for Parkinson's patients, making use of the fact that, like the nerve cells that die in the disease, cells from the adrenal glands also secrete dopamine. Backlund grafted tissue from the patients' own adrenals into the brains of two patients suffering from severe Parkinson's and found some minor but transient recovery of function.[13]

Then, in 1987, in a way that oddly prefigured the Hwang affair twenty years later, the field was blown wide open by a dramatic report from a hitherto little known group led by Ignacio Madrazo at the La Raza hospital in Mexico City, who scooped the Swedes by claiming to have dramatically improved movement, speech and other symptoms in two Parkinson's patients by transplanting tissue from their own adrenal glands.[14] Six months later, Madrazo went further, stating that he could now restore function by using tissue from a spontaneously aborted foetus – that is, no longer using an autologous transplant. Videotapes of these patients were so sensational that Madrazo was received with huge applause at clinical and scientific meetings. Overnight, he became a celebrity, besieged by the press and with Parkinson's sufferers from across the world writing to him requesting treatment. 'Either these results are erroneous in some way, or they are real and dramatic ... there is no reason to question the integrity of the scientists', said Curt Freed, of the University of Colorado.[15]

Hospitals in Sweden, the US and UK hastened to seek approval for similar operations, but within a year doubts had begun to set in. The American Academy of Neurology issued a position statement urging caution, in response to what was described as an 'unseemly haste' by clinical centres in the US rushing to use the foetal transplants. More than a hundred had been performed by March 1988, but most results were negative. The Swedish group at Lund, perhaps the most experienced in the world in such studies, also failed to replicate the Mexican data. Before long the operations were discontinued, and the researchers went back to the drawing board. Huge problems needed to be solved. 'Harvesting' and maintaining the cells, deciding whether cells from several embryos could be pooled, and discovering how to direct and control their growth and differentiation in the brain to ensure that they could survive, grow, replace the dead dopamine-secreting cells, and make new connections without swamping other regions of the brain with unwanted growth, were all issues that remained to be resolved.

Twenty-five years later they are still unresolved. Animal experiments continued, but, following the rise of the new genetic techniques through the 1980s and '90s, with a quite different approach. Now the suggestion was that one might be able to replace the dying dopamine-secreting cells by somatic cell genetic engineering. The aim was to use a modified virus (known as a plasmid vector) into which the DNA coding for the enzymes needed for dopamine synthesis had been inserted and inject it into the brain, in the hope that the foreign DNA would become incorporated into the brain cells and begin churning out the neurotransmitter. Results in animal models suggested that this might prove technically possible, but there were huge problems in ensuring that the engineered DNA was confined to the cells it was intended to transform. As a consequence, by the new millennium the promise of stem cells had once more trumped genes as the biological structures offering clinical hope, and the genetic engineering approach fell out of favour as part of the failed dreams of a past century.

Meanwhile, as John Sladek and Ira Shoulson wondered in a perceptive article in 1987, granted that there were many excellent animal models for Parkinson's disease, why the rush to work on humans?[16] 'Has the animal rights lobby intimidated investigators to the point where they are being unduly pressurized to carry out animal research in humans?' they asked, asserting that 'we should first be persuaded that the immediate scientific and practical questions regarding grafting are unanswerable in animal experiments'. Sladek and Shoulson's question might well be addressed to today's hESC researchers.

BRINGING THE STRANDS TOGETHER: CLONING AND STEM CELLS

There are two major clinical avenues along which stem cell research can be directed. If the problem is one of the repair or replacement of damaged tissue, as in skin grafting or repair of spinal damage, what is required is a supply of autologous pluripotent stem cells. Before Yamanaka's technique of creating iPSCs from skin cells, hESC seemed to offer the best chance. As pluripotency is controlled by certain factors in the cytoplasm, but virtually all the DNA is in the cell's nucleus, the goal was to combine the DNA from a cell taken from an adult patient with the cytoplasm from a pluripotent cell. In practice this means asking a woman to provide ('donate' is the usual term but it is not always a gift but instead a commodity) the egg (or more exactly an oocyte), removing its nucleus, and replacing it with the nucleus from the patient. The cells of the developing blastocyst would be identical to those of the adult cell from which the original nucleus was derived,** and cell lines derived from the blastocyst would thus be clones of that adult and thus effectively autologous. This process of enucleating an oocyte and replacing the nucleus with that of an unrelated adult is called somatic cell nuclear transfer

** Although this isn't absolutely true as the blastocyst would still contain the DNA present outside the oocyte's nucleus – mitochondrial DNA.

or SCNT. SCNT would also realise a second clinical dream by making some forms of genetic engineering possible – for instance replacing a nucleus containing DNA-carrying genetic risk factors with another 'healthy' nucleus. It also makes possible another line of research that has generated intense ethical debate: implanting the nucleus from a human cell into an animal, thus creating human-animal hybrids or chimeras.

The work that made all this possible started with frogs. In 1962, the Oxford biologist John Gurdon destroyed the nucleus of an egg from a claw-toed frog and replaced it with the nucleus of a cell taken from the intestinal lining of a tadpole. The egg grew into a tadpole genetically identical to the tadpole that had provided the implanted nucleus – thus a clone of that tadpole – the first demonstration that it was possible to clone a vertebrate. The egg's cytoplasm had restored the totipotency of the tadpole cell's DNA. Gurdon's work was received with huge excitement by the scientific community. However, replicating his results with other animals was unsuccessful and led to the conclusion that the approach was restricted to frogs. This difficulty of successfully transferring approaches between one species and another – as in the case of Mendel, whose famous ratios found in peas did not occur in several other plant species – bedevils the life sciences. In the case of both scientists, their brilliant achievements seemed initially to be cul-de-sacs.

DOLLY RAISES THE STAKES

It took many false dawns – and another thirty-five years – before Gurdon's work could be replicated in mammals. The first success came in 1997 when the Roslin team of which Ian Wilmutt is the best-known member, announced the birth of Dolly the sheep. Dolly was soon followed by the successful cloning of a variety of other species, including cattle, and, in 2005, a dog, called Snuppy by the Korean team led by the veterinarian Woo Suk Hwang.

The birth of Dolly intensified the argument that had been raging in many countries over the ethical and legal implications of the use of the new techniques for research on human embryos. There was widespread distaste – even revulsion – at the idea of human cloning (though there were some conspicuous exceptions: Richard Dawkins told the BBC in 1999 that he wouldn't object to his daughter being cloned, though what she felt about the prospect was unclear). Because the idea of cloning humans was self-evidently abhorrent, many, including the British government, assumed that it was already illegal. It was only when Catholic activist Josephine Quintavalle challenged this tranquil assumption and was sustained by the court that the government realised it had to legislate. This it did in 2001. Today, in most but by no means all researching countries, human reproductive cloning has been outlawed, and in 2005 the UN passed a non-binding convention banning human reproductive cloning.

Research into cloning opened up the possibility of making specific human cell lines with known genetic deficits, from cloned blastocysts, which could be a valuable research tool for studying disease. Of even greater significance, the researchers argued, was the potential therapeutic value of the technique through SCNT. But cloned hESC could also potentially serve for reproductive cloning. So, in response to the ban, an alliance of stem cell researchers and bioethicists argued that there is a difference in intentionality driving the research. There is, however, no evidence to support the claims made over the years by maverick scientists that they have actually successfully cloned a human. (The therapeutic/reproductive distinction is reminiscent of the relationship between nuclear power and nuclear weaponry. For many years nuclear physicists, supported by politicians, successfully convinced the public that the two were entirely distinct. This hypocrisy is revealed by the West's response to Iran's nuclear ambitions.)

There were still technical problems to overcome. It was James Thomson, a Wisconsin-based veterinarian and molecular biologist, and his colleagues who made the breakthrough. Working in collaboration with gynaecologists from Israel, whose patients provided the donated thirty-six embryos, the team was able to generate five distinct ES immortalised cell lines, three male, two female, using procedures developed first in mice and later in non-human primates.[17] The authors had no doubt of the potential significance of their work, pointing to the potential for drug discovery and treatment of diseases such as Parkinson's. *Science*, in which the paper was published, also provided a commentary by the gynaecologist and fellow stem cell researcher John Gearhart (who in 2009 was invited to the White House to celebrate Obama's overturning of the Bush ban):

> It is exciting to speculate on how human ES lines could be used ... Obvious clinical targets would include neurodegenerative disorders, diabetes, spinal cord injury ... In addition to possibly providing large numbers of pure populations of cells for transplantation, ES cells would also lend themselves to several strategies for the prevention of immunological tissue rejection after transplantation, including ... creation of universal donor lines ... production of ES lines containing the genome of the prospective recipient ... it may eventually be possible to do nuclear transfers into pluripotent stem cells, which could then be expanded and differentiated.[18]

THE HWANG AFFAIR: EUPHORIA, HYPERBOLE AND FRAUD

Thomson's human cells were grown on a bed of 'feeder cells' derived from mice, so were not immediately suitable for therapeutic use, and problems of immune rejection of transplanted cells remained to be solved. Six years later, however, the promises made by Thomson and his team seemed to have moved dramatically closer to being realised. In 2004 Hwang's team claimed to have generated human embryonic cells from cloned blastocysts using SCNT and, importantly, grown on human rather than mouse feeder cells.[19] *Science*'s staff journalist Gretchen Vogler in its online prepublication drew attention to the achievement as being both 'remarkable and inevitable'; the South Koreans (once again a team including veterinarians and gynaecologists) had pulled off two firsts, both generating human embryonic stem cells and also demonstrating that they could be used for cloning.[20] To both surprise and acclaim from researchers and press alike, Hwang seemed to have at last achieved in humans what Gurdon had first done in frogs, taking 242 oocytes from sixteen donors and using them to develop the procedures necessary for successful SCNT. In this, the first of Hwang's two reports, nuclei were removed from the oocytes, and then transferred back into the same oocyte to check that the procedures worked, and that cell lines could be prepared from the resulting blastocysts.

Hwang's breakthrough article was published as hard copy the following month, heralded by a rapturous editorial by *Science*'s chief editor, the environmental scientist Donald Kennedy, entitled 'Stem Cells Redux'.[21] Just over a year later the Hwang team published a second, even more dramatic paper.[22] Working with a further 185 donated oocytes from eleven women, and implanting the enucleated eggs with nuclei taken from skin cells of biologically unrelated individuals carrying genetic diseases such as immunodeficiency, or juvenile diabetes, or spinal cord injury, Hwang claimed to have created patient-specific embryonic stem cell lines, potentially available for either research or therapeutic cloning. Seemingly at a stroke, the formidable technical problems had been solved; the promise of hESC was at last about to be realised.

The response by the stem cell research community, and in Britain the regulator, was immediate and positive. Suzi Leather, Chair of the UK HFEA, issued a press release referring to it as 'an exciting advance from a reputable group in South Korea. The scientists are trying to develop new therapies for degenerative diseases ... It is a responsible use of technology and an important area of medical research.'[23] Not to be outdone, the Newcastle team of Alison Murdoch and Miodrag Stojkovic welcomed Hwang's success, acknowledged that the Koreans now led the field, but also announced that they too had successfully produced a blastocyst. Britain was thus still in the game. Scotland, home of the Roslin Institute, was not to be left out of the action and issued a press release announcing a joint Scottish–Korean collaborative seminar that Hwang would address.

The UK Science Media Centre – which describes itself as independent, 'unashamedly pro-science', and as bringing the voices of the scientific community to the public – invited comments from stem cell scientists, a bioethicist and two CEOs of patients' organisations.[24] Thus Linda Kelly, of the Parkinson's Disease Society (now Parkinson's UK), and Alastair Kent, director of the Genetic Interest Group, a large umbrella group for patients' organisations were invited to contribute. (Kent's close association with the biotech industry has long been criticised by Lobbywatch.) Both extended a warm welcome to the Hwang breakthrough. More critical voices, such as those from the disability movement and Hands Off Our Ovaries, were not invited.

Roger Pedersen, Cambridge Professor of Regenerative Medicine, praised Hwang as 'substantially advancing the cause of generating transplantable tissues that exactly match the patient's own immune system'. John Sinden, Chief Scientific Officer at the biotech company ReNeuron, began his welcome with the words: 'Ever since Dolly the sheep was created in 1996, the scientists have been pursuing the Holy Grail of determining if embryo cloning is possible in humans.' With this rhetorical move, the scientist appropriates both the language of Creationism and that of the Arthurian Christian knights to celebrate the project of the human embryonic stem cell researchers. Invoking the Grail reprises the earlier claims by Gilbert for the HGP. But Sinden, having mobilised Christianity-soaked metaphors around human reproduction (scarcely a reassuring move), went on to offer reassurance, saying that 'This is not to raise the spectre of reproducing cloned babies, but rather to generate what would be a remarkable therapeutic solution to a wide range of intractable diseases.' Here Sinden raises high hopes for those with such diseases, not confining himself to the hope for more effective therapeutics but offering nothing less than a 'solution'. Having raised the cultural stakes with his hyperbolic language, Sinden concluded by toning down the euphoria he himself had generated by adding that 'there is no therapeutic hope on the early horizon'.

The media furore surrounding Hwang and his colleagues was comparable (though very different in national style) to that following the theatrical presentation of the 'first draft' of the human genome in 2000 by Clinton, Francis Collins and Craig Venter on one side of the Atlantic and Blair on the other, which had offered stereotypical and complementary nationalisms: go-getting wealth-creating genomics in the US; public-interest genomics in the UK. In South Korea, Hwang became a national hero overnight, surrounded by adulatory crowds, featured on a postage stamp, offered a lifetime of free flights by South Korea's national airline. Glowing in the media warmth, he announced that he proposed to make his country into a 'world-wide hub of hESC research'. *Time* magazine named Hwang as one of the world's most influential people. US researchers meanwhile ruefully speculated that the Bush ban had resulted in an irretrievable loss of research leadership to the new Asian tigers, unhampered by ethical restraints (though in fact South Korea had established ethical guidelines for stem cell research a couple of years earlier).

But how had Hwang's project succeeded where others had failed? He himself suggested that it was based on a difference between his and others' techniques of extracting the nuclei. While the stem cell researchers greeted the work enthusiastically they were openly envious of the sheer number of eggs available to Hwang's team. *Science* quotes stem cell researcher Jose Cibelli of Michigan State University as saying: 'More than 200 eggs? Wow, I'm drooling.'[25] It was, Hwang suggested, the abundance of eggs and the 'magic hands' of his team that had made possible various modifications to their procedures, eventually leading to success in getting blastocysts to develop into cell lines. Both *Science* and *Nature* reported that Hwang's team of dedicated women technicians were workaholics, working a seven-day week starting at 6.30 every morning except Sundays, when they started at 8. Hwang was reported to be a Buddhist, getting up regularly at 4.30 to pray. The media raised the spectre of this successful therapeutic cloning opening the door to reproductive cloning, although in South Korea the latter is illegal. Amid the media storm, only US stem cell researcher Rudolf Jaenisch publicly questioned Hwang's claims.

Even while South Korea was celebrating, a whistle-blower was suggesting to the press that far from the women freely donating their eggs, Hwang had paid them. Within a month of Hwang's first paper in February 2004, the Korean Bioethics Association convened a committee to consider the issue. They posed the question of whether possible vulnerable donors 'maybe from young female researchers or women with a conflict of interest?' Judging by what was to follow, the Bioethics Association had its ear close to the ground, but in view of the public rejoicing they had to proceed with great caution. But proceed they did, and in 2005 a law was introduced requiring a licence to research on human eggs. Meanwhile, outside South Korea, the main scientific journals began to pick up on these ethical concerns. In May 2004, a *Nature* journalist interviewed a young PhD student from Hwang's team who claimed that she had donated eggs in a spirit of altruism and patriotism. After Hwang denied that she had been a donor, the student retracted her account, blaming her weak English for the misunderstanding.[26]

Five months after the dramatic 2005 paper, Hwang admitted that the 'donors' had been paid (illegal in South Korea), and also that he had used eggs from women members of his research team. He claimed that he had lied about the researchers' donations to protect their privacy. As the story unravelled, Hwang's gynaecologist colleague Sun il Roh, who shared the patent, confessed that he had paid twenty women for eggs at the rate of 1,430 US dollars each – 'not a large amount of money considering that they had to receive injections every day for 8–10 days'.[27] In much of the US, selling and buying human eggs is legal, mostly for IVF but also for stem cell research – as Cibelli himself had done. In Europe payment is illegal but 'compensation' acceptable. In Britain the HFEA has set the figure at £750 – very similar to the sum Hwang had paid – a similarity which must raise questions about the difference between payment and compensation. Nonetheless the Korean bioethics community remained troubled, and Gerald Schatten, the single American co-author on the paper, announced he was breaking off relations with Hwang and withdrawing his name.

Bad as this was for Hwang's legitimacy, what was to follow was worse: the collapse of the credibility of the science itself. The claim that he had established distinct patient cell lines was shown to be false by independent tests commissioned by a Seoul-based television company. Closer scrutiny of some of the images of cells in the paper were revealed as fraudulent, copied from earlier papers.

A letter from eight stem cell researchers called for independent verification of Hwang's claims. Finally, following the report of the investigation committee of Seoul National University in December 2005, the papers – which, it must be recalled, had gone through the rigorous reviewing process that *Science* demands – were formally retracted. Despite the focus of the Western press on the nationalist fervour around Hwang, the Korean academic community behaved in an exemplary fashion. Meanwhile Hwang, stripped of his titles, has continued to maintain that his results were essentially valid, and vowed to continue his research.

CHIMERAS: TRADING ONE ETHICAL PROBLEM FOR ANOTHER

As we have said, there have been two strands of ethical concern over the provision and the uses of hESC: the feminist critique of the mining of women's bodies to provide a research resource; and the religious and human dignity critique of the use of human embryos. It wasn't long before the research community came up with a proposal that sidestepped the first, even while it confronted the second with a further, unanticipated challenge. If the capacity for pluripotency resides in the cytoplasm of an embryonic cell, while the major proportion of the genetic material – the DNA – lies in the nucleus, it would be possible to take the nucleus of any adult human cell – say a skin cell, and implant it in an animal oocyte by SCNT. The resulting blastocysts should then express human genes – they would be human-animal hybrids (sometimes called chimeras, although technically a chimera is an organism containing a patchwork mixture of human and animal cells). This meant that hESCs would no longer be required, thus allaying feminist criticism.

But what would be the ethical status of these hybrids? From their very beginning, the techniques of genetic manipulation have been employed to create such hybrid forms, although, as we discussed in Chapter 3, the concerns initially raised were about safety rather than ethics. Incorporating the gene for human insulin into E. coli, which could then be grown in quantity, and the insulin harvested, presented manufacturing but not ethical problems. Humulin, as the product was named, was developed by Genentech in 1978 and has been bulk produced and marketed by the pharmaceutical company Eli Lilly ever since. Animal models for human diseases – for instance mice into which genes implicated in Alzheimer's disease have been inserted – have become standard laboratory tools. However in the post-Dolly era, when the idea of inserting genes into cloned cattle to manufacture human proteins was mooted, bioethicists began to speak disparagingly of 'the yuk factor' as an irrational obstacle to research. By naming it irrational the bioethicists failed to explore 'yuk' as the expression of an as yet unarticulable ethical response.

The prospect of growing human-animal chimeras in which a large portion of human DNA would be expressed ratcheted up the temperature. Stanford bioethicists debated a thought experiment proposed by lawyer Hank Greely. Imagine taking a great ape and replacing the genes responsible for building its brain one by one with their human equivalents such as the gene *FOXP2*, which has been suggested as key to the human capacity for speech. At which point would the ape become a human? After 10 per cent? 30? 50? Given that that chimpanzees and humans share more than 98 per cent of their genes, it is difficult to see what might be gained by such a thought experiment. However, improbable though such procedures would be, such speculations began to occupy the pages of the bioethics journals.

In 2007, the HFEA began a public consultation over the use of human-animal hybrids and chimeras. At that stage Federal funding for hESC research in general was still blocked in the US, and making chimeras specifically prohibited in Canada, Australia, and several European countries. Respondents to the HFEA's opinion polls and participants in open public meetings were opposed to such research, though tended to change their opinion when their 'misinformation' had been clarified. By contrast, the scientists consulted were overwhelmingly in favour of research going ahead on all fronts. (Stephen Minger described how he had spent much time in Westminster lobbying for the hybrids.) After careful consideration, the HFEA concluded that research, under licence, should be permitted.[28] In 2008, legislation to update the 1990 Human Fertilisation and Embryology Act was introduced. A cross-party group of MPs, supported by Catholic members of the Labour Cabinet such as Ruth Kelly (close to Opus Dei) and the majority of the Tory Shadow Cabinet, moved amendments to ban the creation of hybrids, along with 'saviour siblings'. In a free vote, the amendments were rejected (then Leader of the Opposition David Cameron and PM Gordon Brown both voted to permit the hybrids.)

Three years later a working group from the Academy of Medical Sciences revisited the issue,[29] by which time researchers in the UK had received licences for work on some 150 hybrids or chimeras. The Academy is a relatively new institution, established in 1998 with one its key aims being to link academia with industry. Apart from its scientific fellows, honorary fellows include Sir David Cooksey and Lord Sainsbury. A further public consultation exercise and opinion poll found the public now 'broadly supportive' of research using animals containing human material. For many respondents the concern was more with animal welfare than with human dignity, except when the chimeras involved cells or genes specific to human brains or reproductive systems. The Academy concluded that any such proposals should be subject to consideration by a special panel. Chimeras and hybrid embryos between human and non-human primates should not be permitted to survive beyond the fourteen days laid down in the original Warnock Report.

iPSCs TO THE RESCUE?

The confidence of the stem cell researchers, and even more of their advocates in industry and government, thus soon recovered after having been briefly dented by the fallout from the Hwang affair. Hwang was the bad apple in the barrel, and there was more than a hint of racism in the suggestion that this was an Asian phenomenon that couldn't happen in the West. This ignored the fact that the intense competition had already made fraud an endemic problem in the life sciences, much discussed by scientific journal editors in both the US and Europe. The community would, it hoped, police itself through the International Society for Stem Cell Research

(ISSCR) – a predominantly US organisation – and tougher reviewing procedures by the lead journals. Fortunately for the field, within the year unexpected salvation appeared in the form of Yamanaka's iPSCs. First in mice, and then in humans, his group succeeded in 'reprogramming' adult human skin cells, using a retrovirus to insert into their DNA four genes known to be associated with embryonic development. The transformed cells became pluripotent.

A whole new world opened up. There could be no ethical problem with using human skin cells, a virtually limitless source of material. Because a person's own cells could be used, autologous cell lines could be created which ought to prevent possible immunological rejection. Furthermore, taking skin cells from people suffering from genetic diseases would enable the cellular processes and possibly treatments for a disease to be studied in vitro. By 2011, hundreds of such cell lines had been created modelling everything from heart defects to schizophrenia, and the first trials were underway in California to treat the virulent genetic skin disease dystrophic epidermolysis bullosa.

While some stem cell researchers turned to iPSCs with enthusiasm, others who had staked their careers on hESCs were forced onto the defensive in order to justify continuing experimenting with human embryos. After all, they argued, the genetic manipulations required to create iPSCs meant they were not exactly the same as a proper pluripotent embryonic cell, and might even be hazardous, as the retrovirus required for the transformation might also be cancer-inducing. The iPSC researchers responded by modifying Yamanaka's technology, using chemical agents rather than a virus to transform the cells. But new problems continued to appear; the cells, it seems, are not quite pluripotent, and as researchers gain experience with them they discover immunological, genetic and epigenetic abnormalities.[30]

FROM BENCH TO BEDSIDE: STEM CELL TOURISM

As always, the path from bench to bedside is far from simple. Growing hESCs or iPSCs in small numbers in Petri dishes in the controlled conditions of a cell culture lab is one thing, but scaling them up for batch production to provide the billions necessary for any clinical use requires quite other technologies. This is a particular problem with iPSCs because, whether a retrovirus or chemicals are used, only a tiny fraction, less than one in ten thousand of the cells, are made pluripotent. As the transformed cells multiply there are dangers both that they mutate or become carcinogenic, or that the cell lines become contaminated. A Fordist production line, rigorous quality control, techniques for cell sorting, storage and transportation where several tens of different regulatory bodies are involved, make the process too slow for those seeking dramatic quick-fixes.

Even if these problems have been overcome, the crucial question is whether the stem cell therapies will actually work. For some, the answer is clearly yes when autologous cells are used – skin grafts for example, or the remarkable bioengineering achievement in 2011 of growing an artificial trachea to treat a patient with tracheal cancer by seeding a polymer-based scaffold with the patient's own stem cells. There is also hope for rebuilding damaged corneas by using a special subtype of stem cells. But for many of the cures prophesied with such enthusiasm by Yvette Cooper at the millennium, there is still a long road to travel. Parkinson's disease is the paradigm case. The problems of how many cells to inject, how to control their growth and localise it to the proper region of the brain, how to ensure that the new nerve cells make functional contacts with their neighbours – none of these fundamental issues have been resolved in animal experiments despite decades of work.

In a major blow to any thoughts of rapid progress, in 2011, just two years after the start of its trial using stem cells to treat stroke, which had been seen as one of the first fruits of the Obama legislation, Geron announced, to the general dismay of the stem cell community, that it was abandoning the trial and stem cell research in general to focus on cancer. What was to happen to the patients enrolled in the trial was unclear. Hope once again was deferred, this time by Geron's business perspective, which recognised that there would be an insufficient return on its investment.

In a situation where, with few exceptions, stem cell therapies have not moved beyond clinical trial into FDA approval and yet the hype and hope of miracle cures has been loudly proclaimed, the international marketplace provides a solution. For those who can pay, shadowy companies like Florida-based Regenocyte Therapeutics, Beike Biotechnology in China, and Theravitae in Thailand have set up a flourishing trade in stem cell tourism. Others, in the Netherlands and Ireland, have been closed down by regulatory authorities or, like Germany's XCell-Center, filed for insolvency. The companies' websites promise treatments for lung, heart and spinal cord diseases where conventional therapies have failed, and are aglow with euphoric testimonials from patients. For the American or Western European companies, the treatment centres are offshore, in Romania, the Dominican Republic or Thailand. A typical US patient will pay anything from $20,000 upwards to enter into a truly global network, in which extracts of their blood or bone marrow will be shipped to Israel, where regulation is light and skills high, stem cells will be extracted and mixed with growth factors to multiply, and then couriered to a treatment centre in Latin America, Asia or Eastern Europe timed to coincide with the arrival of the patient. The cells will then be injected into the heart, lung or other diseased organ. Improvements are said to begin three to six months after the injections.

Do the treatments work? Leading EuroAmerican stem cell researchers are sceptical. The companies release few details of their procedures, control studies or statistics of their success rate, and what they do supply is rarely peer-reviewed. The painstakingly established stages through which any new drug or treatment is supposed to proceed before being licensed have simply been leapfrogged. The ISSCR has criticised the companies and called for tighter regulation,[31] although it realises it cannot regulate internationally. In this context, Stanford ethicists Charles Murdoch and Christopher Scott have argued that there is an ethical duty to use social media to:

> 1. Disclose and discuss [with prospective patients] the potential for real physical, psychological and economic harm from the

interventions and travel, including costs of the procedure relative to patient's means.
2. Disclose and relay independent scientific evidence of risk or benefit for a defined intervention.
3. Disclose any evidence of ethical misconduct or questionable practices. This includes:
   - Failure to supply local and national evidence of oversight.
   - Engaging in questionable patient recruiting practices.
   - Clear misrepresentation, fraud, or patient abuse.[32]

As with so much bioethical advice in an age of globalisation, however, there is a conspicuous silence over just who is to bell this particular cat.

8

The Irresistible Rise of the Neurotechnosciences

TWO DECADES

The 1990s wasn't only the decade of the human genome. Quietly, without fanfare, the US National Institutes of Health announced that it was also to be the Decade of the Brain. Always a step or two behind the US, it took the EU funding agencies practically to mid-decade before they too adopted the slogan. However, on neither side of the Atlantic was there a clear target: no equivalent in brain terms of the HGP's sequencing of the human genome; no Clinton–Blair press conference at the end of the decade to announce that the brain had been 'solved'. What there was (and probably more significantly for the expanding numbers of neuroscientists) was a massive increase in state, pharmaceutical, foundation – and military – funding. So much so that by the millennium researchers were claiming that neuroscience was now entering its 'Decade of the Mind', and the publishers' lists were awash with books proclaiming neuroessentialism. Yet 2011, a year in which mental and nervous diseases were being diagnosed at record levels, was also the year in which the largest pharmaceutical companies researching, developing and marketing psychotropic drugs declared that they were withdrawing from the field, to focus on more tractable diseases, in particular cancer, where gains were evident.

This chapter charts these two decades, and asks what have been the drivers pushing the neurosciences forward? For the neuroscientists themselves, it has been the pursuit of biology's last frontier – to understand the human brain, and, through the brain, the mind itself. For psychiatry, advances in genomics and the neurosciences powered up the promise it had made in the 1950s to provide the scientific foundations its practictioners had long sought to give them equal status with the physicians and surgeons. Since the birth of this new biological psychiatry in the '50s the pharmaceutical industry has been integral to the process. As mental illness became both more prevalent and increasingly costly, the new drugs promised patients therapies of unprecedented power and governments a huge potential for wealth creation. Today, even as Big Pharma has recognised that neuropharmacology is not fulfilling its promise, hitherto undreamed of neural devices are being introduced to the market by newly formed neurotechnology companies, hybrids – like those of biotech – between academia and industry.

By the end of the NIH's Brain Decade, the annual meetings of the American Society for Neuroscience (ASN) were attracting international attendances of 35,000 (the European equivalent never got beyond 8,000), drawn not just from university research laboratories and pharmaceutical companies but by reps from the newly formed neurotech companies as well as somewhat shadowy military representatives. (Though such has been the interlocking of these sectors that it was possible to find high-flying researchers with their feet, hands, salaries and research funding in all four camps at once.) Neuroscience, now transmuting into a fully fledged technoscience, continued to expand through the following decade, appearing prominently alongside genomics in the priorities of national and international funding agencies and the massive EU Framework Programme. Scoping exercises such as the Foresight Report in the UK saw the field as a major growth area, ripe for wealth creation. In the US, the National Research Council highlighted the potential of cognitive neuroscience and its related technologies, though a bill to establish a national neurotechnology initiative was sidelined by Congress. The neurotech industry was becoming more organised, and by 2010 Zack Lynch, founder of the US-based Neurotechnology Industry Organisation was claiming to speak on behalf of more than a hundred companies. In the same year Neurotech Business Reports calculated that the sector had a turnover of $4.3 billion, forecast to rise to $10.2 billion by 2014.[1]

With the Decade of the Mind, the neuroscientific claims became ever more extravagant, going way beyond those of Koshland at the inauguration of the HGP that it would solve everything from schizophrenia to homelessness, or even of the *Guardian* editorialist that stem cells would enable the lame to walk and the blind to see. Neuroscience, its advocates believed, would not merely conquer psychiatric and neurological disease, it would tackle the last great mystery of life that neo-Darwinism had not resolved – that of human consciousness. It would explain morality, selfhood and the source of religious experience, and provide a new framework into which romantic love, ethics, law and even aesthetic experience and literary criticism could be given a brain location. A new internal phrenology appeared to be in the making.

Among leading neuroscientists, there was no longer a mind-brain problem. For Francis Crick 'You are nothing but a bunch of neurons.'[2] Nobel Prize winner Eric Kandel – sometime psychoanalyst but now arch reductionist memory researcher whose model organism is the Californian sea-slug – similarly claims 'you are your brain'.[3] For neurophilosophers such as Patricia Churchland,[4] the reductionist agenda is clear; the language of what they disparagingly refer to as folk psychology is to be swept away and replaced by the precision of neuro-computation. In a not dissimilar vein, the psychoanalyst Mark Solms suggests that neuroscience has advanced so far that it is time to return to Freud's original, biological project of locating psychic problems in brain processes.[5] Even the Dalai Lama has enthusiastically welcomed the efforts of his neuro-minded American followers to demonstrate the efficacy of meditation by imaging the brain processes involved. For many philosophers and ethicists, advances in the neurosciences have raised new problems and with these the possibility of a new field (or perhaps that should again be enterprise) of neuroethics. If the neurosciences had shown that minds were no more than the epiphenomenal product of brain processes, what becomes of the concept of free will and autonomy? Both ethicists and neuroscientists continue to frame the question in the language of traditional philosophy rather than the newer, sociologically grounded, concept of human agency.

If human actions are the product of determinate brain processes, what remains of the legal concept of responsibility? Some neuroscientists are happy with what one contentedly refers to as the 'ruthless reductionism' that relegates consciousness to molecular processes in specific regions of the cerebral cortex. Others have fudged the issue, much as Dawkins had done with the tyranny of the selfish gene. Thus for Gerald Edelman, Nobel immunologist turned consciousness researcher, 'You are your brain – plus free-will.'[6] Neurophysiologist Benjamin Libet, having demonstrated that decisions to act are made in a person's brain several tenths of a second before they state they have consciously made that decision, reaches for a sky-hook with his proposal that although free will does not exist, a person retains a 'free-won't' located in some conscious mental field, able to countermand the brain's decisions.[7]

With the mind–body problem apparently solved, the neurosciences were confident that with more resources for more research they could find new and effective therapies for the rapidly increasing numbers of those diagnosed with mental and neurological diseases, from depression to Alzheimer's disease – this latter already affecting 700,000 in the UK alone and projected to double by 2020 as populations aged. The WHO announced a pandemic of depressive illness across the globe, and by 2011 reported that mental and nervous system disorders constituted 13 per cent of the 'global disease burden'. Despite Big Pharma's best efforts, the world, it seems, has been getting mentally sicker. Having been a rare diagnosis three decades previously, by the millennium one in forty Americans was classified as suffering from bipolar disorder. In Britain, Attention Deficit Hyperactivity Disorder (ADHD), a condition afflicting perhaps one in 500 children in the 1980s, is now estimated to be twenty-five times more common. In 2011, the European Brain Council calculated that in Europe 38 per cent of the population – 165 million people – would develop mental illness in any one year at an annual cost of €800 billion or 24 per cent of the Continent's health-care budget, although they conceded that many of these 165 millions would go unrecognised and untreated.[8]

The tonnage of psychotropic drugs ingested has continued to

increase year on year. Antidepressant prescriptions rose by 28 per cent in the UK between 2008 and 2011, from 34 to 43.4 million; anxiety reducers from 6 to 6.5 million and sleeping pills from 9.9 to 10.2, perhaps in response to the intensifying precarity produced by the worsening economic crisis. Prescriptions for ADHD increased from 2,000 a year in the early 1990s to over 600,000 by 2010. In the same year the global market for mental health related drugs was some $75 billion. Not that all this benefited Big Pharma, as many of the drugs were running out of patent; cheaper generics, often manufactured in India, were out-competing them and there were no new potential blockbuster drugs in the pipeline. For that matter, many in the West have turned to herbal therapy, particularly to St John's Wort for mild to moderate depression and anxiety – a self-treatment reported as having fewer unwanted effects than most drugs produced by Pharma. Meanwhile patient groups, such as the Schizophrenia Association and the Alzheimer's Society, protested against the injustice that the proportion of the biomedical research budget being spent on mental health and brain was much less than that for cancer and heart disease. The President of the British Neuroscience Association pointed out the anomaly that a third of the UK's health-care budget goes on the troubled mind and the hurt brain, yet the neurosciences receive only one sixth of the biomedical research budget. The situation is similar in the US.

Nonetheless it was a shock when several of the biggest pharmaceutical companies announced, separately but almost in lock-step, that they were abandoning attempts to develop new drugs to treat psychiatric conditions such as schizophrenia and depression, closing laboratories and sacking research staff in order to refocus their research on health problems that lay less ambiguously within the body. These moves have to be set in the context of the turbulent history of the pharmaceutical industry over the past two decades, during which it had been transformed simultaneously in two contrary directions. On the one hand analysts argued that there were too many small companies, and that the future lay in mergers to create a small number of global players. Thus, from the 1980s onwards, Smith Kline and French became Smith Kline Beecham,

Glaxo became Glaxo-Wellcome, and finally the two coalesced into GlaxoSmithKline; Rhone-Poulenc merged with Hoechst and then Sanofi-Aventis; ICI merged its drugs division with the Swedish company Astra to become AstraZeneca; Ciba became CibaGeigy and then merged with Sandoz to become Novartis. But as the companies became larger they also began to outsource research to the newly created small biotech companies, just as Roche had done with Genentech and deCode. Some of these smaller companies, faster on their feet than the giants, were, like Genentech, already focused on cancer where so many gains had been made.

The bellwether of the 2011 withdrawal from psychotropic research was Pfizer, still the world's largest among the Big Pharma, which closed its major research complex at Sandwich in Kent with the loss of 2,500 jobs. From New York to Oxford, start-up biotech companies with eminent neuroscientists as their CEOs or chief scientific advisers, financed by venture capital on the promise that they had the key to developing drugs to enhance or erase memories or to treat Alzheimer's disease, went belly-up. Despite strong UK government pressure, Novartis closed its manufacturing base and part of its R&D facility in Horsham with a loss of 550 jobs, as well as its neuroscience facility in Basel. GlaxoSmithKline and AstraZeneca were not far behind. (With rather less publicity, several of these companies have in fact simply relocated. GSK is researching neurodegeneration at its new sequencing centre in Shanghai; Novartis is opening a new research division to study psychiatric genetics in Cambridge, Massachusetts, and the EU has put up €25 million for a public-private partnership to work on the GWAS data from deCode's schizophrenia study that we discussed in Chapter 6.) In response, the government announced a £100 million boost for four research campuses, including life science centres near Cambridge and Norwich – small compensation for the loss of skilled jobs and the government's previously announced £1.4 billion cut to the science budget.

Meanwhile, the MRC, as part of its translational research strategy, sealed a special relationship with AstraZeneca, ring-fencing £10 million funding for researchers wishing to study novel functions from a portfolio of the company's drugs. The gradual transformation over the past decades in the relationship between academic science and industry from one of relative autonomy to one of fusion and hybridity has, with this project, come to fruition. The problem – even if the project is successful and does cross the gap between bench and bedside – is whether AstraZeneca continues to locate its research and production in Britain. There are good reasons to be cautious; Big Pharma has always been footloose, moving from country to country weighing up the attractions of each. In an age of globalisation the problem is keeping the benefits within the subsidising nation. (One way of achieving this is through 'income contingent loans' where the state invests directly in particular companies, reaping a dividend if and when the product is successful. Thus while Germany as a more competent capitalist country ties investment to the national economy in this way, Britain merely subsidises global Pharma.)

NEUROSCIENCE OR NEUROSCIENCES?

Despite its name, its numerous national societies and its international journals, neuroscience does not exist as a single science in the way that genomics does. The term was invented in the 1960s by scientific visionaries, key among them the MIT neurophysiologist Francis Schmidt, in a deliberate move to bring together the many different ways in which the brain and nervous system were being studied – from neurology, psychology and pharmacology through physiology and anatomy to molecular biology and genetics.

Two enabling technologies above all have helped drive the neurosciences forward over recent decades: genomics and informatics, one approaching the brain from below, the other from above. From below, advances in human genomics opened the prospect of hunting not just for disease genes but also for those associated with such human attributes as temperament, anxiety and self-confidence. Genes can be knocked in or knocked out of mice, more or less at will, to construct animal models intended to mimic human psychiatric and neurological disorders. Molecular neuroscience is charting the intricate interactions of the complex proteins that self-assemble to form the hundred billion neurons and hundred trillion connections between them which constitute the adult human brain. Extraordinary advances in microscopy allow researchers to directly observe the dynamics of the growth and retraction of these connections and even the passage of signalling molecules between them in the living brain itself.

From above, the windows into the human brain opened up by functional magnetic imaging (fMRI), positron emission tomography (PET) and magneto-encephalography (MEG) can be exploited not merely to locate diseased or damaged brain regions but to detect changes in brain activity in people carrying out such varied tasks as solving a maths problem or viewing the photograph of someone they love. Massive strides in information technology are encouraging ambitious researchers to attempt to build a complete model of the human brain. Despite the scepticism of other neuroscientists, Henry Markham in Lausanne has persuaded IBM to fund the Blue Brain Project, a supercomputer designed to do just that. More practically, information technology has begun to make possible direct two-way communication between brain and computer. For someone who is paralysed, electrodes implanted in the motor regions of the brain can signal neural messages to a computer. In turn the computer can control a robot to carry out the thought-wishes of the person, thus bypassing the seemingly intractable biological problems of paralysis that stem cell therapy is still unable to solve.

But the new technologies have done little to resolve deep-rooted theoretical problems. Old binaries (neurological/psychological; nature/culture; cognition/emotion), deriving from long-standing philosophical debates within the history of Western thought, still obstruct new thinking. Approaches to the brain still reflect the problematics and methodologies of the parent disciplines that have gathered together under the prestigious umbrella of the neurosciences. There is still no real integration, no theory of the brain to match Crick's Central Dogma or the Darwinian logic of natural selection. Unless assembled in some vast stadium to listen to a Nobelist grandee in plenary, the 35,000 who meet at the

ASN convene in multiple specialist sections with little connection between them. Even when ostensibly studying the same subject, such as memory, cognitive psychologists, molecular biologists and brain imagers have few points of contact, working with different understandings of the phenomenon they study even when using the same words. Two textbooks on memory, for example, each written by a leading expert in the field, one a molecular biologist, the other a cognitive psychologist, share almost no references in common.[9] For the former, short-term memory is a process lasting several hours during which the memory becomes fixed through altered connections between neurons; for the latter it is a transient fluctuating phenomenon lasting seconds, which the brain mobilises to carry out specific tasks.

fMRI AND SOCIAL NEUROSCIENCE

A fundamental methodological problem for the neurosciences is that the unit of analysis is individual, yet humans are social animals with brains that have evolved to enable their owners to survive in complex social environments, not to solve abstract problems. This requires that people can recognise others as having intentions, emotions and desires analogous to their own – a 'theory of mind' said to be lacking in brain-damaged and autistic individuals. The almost accidental discovery by neurophysiologists that there are neurons in apes, and brain regions in humans, that respond not merely when an individual performs a particular action but also when he or she observes another performing a similar action (mirror neurons) opened the way to a new field of social neuroscience in humans, studying the brain events involved in recognising and responding to others' feelings and emotions. These are the studies that regularly catch media attention, with the false-colour images dramatically highlighting brain regions active when subjects see pictures of their lover, or compete or cooperate with others in gaming or making investment decisions.

While recognising the importance of trying to move beyond the individual into the social, there is a danger of such studies being over-interpreted. The problems arise at two levels, one technical, the other conceptual. fMRI, the technique used in such social neuroscience studies, measures the flow of oxygen-rich blood in small regions of the brain. The first problem stems from the limitations of the technique. At it best, fMRI averages activity over about two seconds in a small block of brain tissue about 0.5mm across. Although this block of tissue seems tiny, so complex is the brain that the tiny block contains some 5.5 million neurons, 22 kilometres of their dendrites, and up to 55 billion connections. Two seconds is a long time when neurons operate on a millisecond time scale, and as the cells use energy either to excite or inhibit their signalling partners, any change in blood flow might mean either excitation or inhibition. And the problems only begin here. The dramatic false-colour images that grace the journal articles and media reports of fMRI studies are the result of extensive mathematical transformations of the raw observations of blood flow and a priori assumptions about which regions to study and how to select sample sizes. One critical analysis that rocked the social neuroscience community examined fifty-four of its most prestigious papers and described them as voodoo correlations, showing 'implausibly high correlations through statistically inappropriate selection of data' (despite the authors being persuaded to change their original title to something more anodyne, the damage was done[10]). Worse still, a group of students reported getting apparently meaningful fMRI images from a dead salmon.[11] Even kind critics of the fMRI studies argue that at best they are mere cartography – providing maps but no explanations of causative mechanisms.

For us, however, the most damaging criticisms of such studies are not technical but conceptual. Even when they attempt to be at their most social, their methodology is essentialist. The complexities of lived experience are reduced to toy problems amenable to experimental manipulation but in the process emptied of real-life reference, as in Mark Hauser's moral dilemmas discussed in Chapter 2, whose neuroscience of ethics was eclipsed by his lapses in the ethics of neuroscience. Take, for example, a paper by Hsu et al. published in *Science* in 2008, entitled 'The Right and the

Good: Distributive Justice and the Neural Encoding of Equity and Efficiency'.[12] The researchers studied the fMRI responses of subjects who were asked to imagine themselves distributing food aid to starving children in a Ugandan orphanage. There is not enough food to go round, so the subjects have to choose whether to share it equally but inadequately (equity) or to give enough food to a selected few (efficiency). Apparently, choosing equity involves one brain region, the insula, while another, the putamen, encodes efficiency. But aid workers rarely confront such simple binaries, their training in triage directs them towards saving as many as possible in horrific and horrifically complex situations. Asking subjects to make such choices while lying alone and safe in the warm if noisy tunnel of the fMRI machine is so abstracted from the reality of being an aid worker in a famine in a violence-torn country that it is hard to take the simple-minded interpretations that the researchers give to their data seriously.

Despite this critique, viewed most optimistically the non-invasive technologies of neuroimaging with its metaphor of windows into the living brain has opened the possibility of crossing the epistemological threshold between the neuroscience of the individual organism – whether Kandel's sea-slug, the Alzheimer mouse or the human animal – towards a neuroscience able to think fruitfully about the profound sociality of the human animal.

THE TROUBLED MIND AND THE DAMAGED BRAIN

Until the 1950s, biologically oriented psychiatrists had few treatments at their disposal, and those they did have were brutal, unsubtle and non-specific – electroshock, metrazol convulsions and insulin coma, and, as a last resort, prefrontal lobotomy. The invention of psychotropic drugs transformed the situation; by targeting specific biochemical systems in the brain they seemed to offer not merely a treatment but an explanation for the condition they were treating. The problem however was – and remains – that there is no obvious biochemical or physiological marker for psychic distress, no physical measure such as blood pressure or elevated blood sugar, no clear abnormality in the brain. If only such a biomarker could be found, psychiatrists would then be able to move beyond diagnosis based on observed behaviour and phenomenological descriptions, and rely instead on a biochemical test. Ideally, it would be a simple colour change, like the clinistix long available to indicate elevated blood sugar for suspected diabetes. For a while, psychiatrists thought that a clinistix equivalent – a test for the activity of enzymes involved in serotonin metabolism in blood platelets – would diagnose depression. But the hope soon faded. Decades spent trying to identify peculiar substances in the blood or urine of people diagnosed with schizophrenia regularly generated elusive artefacts, like the metabolite found in the urine which turned out to be a product of the excessive quantities of tea psychiatric patients drank in hospital. Biological psychiatry's 'schizophrenic urine' had a short reductionist life.

In the absence of such physical measures, the American Psychiatric Association (APA) developed a psychiatrists' 'bible' – the Diagnostic and Statistical Manual (DSM) – from the sales of which the APA makes a handsome profit.[13] First published in 1952 and currently going through its fifth revision, it is essentially a catalogue of reported signs and symptoms which form the basis of the classification of mental and nervous system diseases, categories often influenced by the raced and gendered values of the psychiatrists themselves. Not infrequently this resulted in inappropriate diagnosis and prescription. One conspicuous example was the classification of women experiencing the menopause as pathologically anxious and depressed, leading to the widespread over-prescription of diazepam, an addictive drug. Homosexuality, originally listed by DSM as a disorder, was only declassified in 1973 with the rise of the gay and lesbian movements, and removed from subsequent DSM editions. Old disorders disappear or are renamed. Minimal Brain Dysfunction becomes Attention Deficit Hyperactivity Disorder; Multiple Personality Disorder becomes Dissociative Identity Disorder; Manic-Depression becomes Bipolar Disorder. New diagnoses such as Panic Disorder and Post-Traumatic Stress Disorder appear. Depending on which boxes are ticked, a diagnosis is made and a drug prescribed. The US origins of the manual lie not only in the expanding categories developed through psychiatric research but in the requirements of an intensely marketised medical system in which clinicians can only provide treatment if the symptoms presented to them are classified as fundable by medical insurance.

The history of the earliest psychotropic, chlorpromazine (Largactil in the UK, Thorazine in the US), is typical of many. It is a derivative of a compound originally synthesised by the French company Rhone-Poulonc as an antihistamine, and almost by accident shown to have calming effects on institutionalised patients. Its effects were said to range from 'allaying ... the restlessness of senile dementia, the agitation of involuntary melancholia, the excitement of hypomania' to 'detachment from delusions and hallucinations', to quote from a standard teaching text on psychological medicine from the 1960s. By 1954 the drug was licensed to SmithKline and French and used to treat schizophrenia and mania; within the decade it had been administered to 50 million patients worldwide. Only later was chlorpromazine recognised as producing lasting brain damage and a severe movement disorder, tardive dyskinesia, known by mental patients and psychiatric staff alike as the largactil shuffle. But its use had ushered in a new era of psychotropics in which the task, as one biological psychiatrist put it at a US parents' meeting (where Steven was also a speaker), had become to understand how 'a disordered molecule causes a diseased mind'.

Tranquilisers, antidepressants and anxiolytics soon followed. Some were discovered almost accidentally by chemists tinkering in the lab, as LSD had been. Others were the results of pharmaceutical biopiracy, exploiting the knowledge of non-industrialised peoples concerning the therapeutic effects of certain plants. The antipsychotic Reserpine, derived from the Rauwolfia shrub that had been part of the medical armentarium of Aryuvedic medicine, is among the best-known examples.

As with chlorpromazine, most Western psychotropics were discovered empirically; there was no theory of how they might work. In the usual biomedical style, to find out how, pharmacologists tested the effects of the drugs in laboratory animals, mainly rats. It turned out that many of the drugs interfered with the functioning of neurotransmitters, the molecules that carry the signals from one nerve cell to another in the brain. The simplistic conclusion was therefore that the psychiatric disorder itself was the result of a malfunctioning of the transmitter systems, which the drugs then rectified. Depression, for instance, might be caused by an excess or lack of such transmission in key brain regions. To test this further, researchers began to develop animal models that mimicked in some way the condition in humans. This is not too difficult for straightforwardly neurological diseases. Cutting off the blood supply to regions of the rat brain results in the equivalent of a stroke; destroying the neurons in the substantia nigra mimics Parkinsonism. Making rats depressed, anxious or schizophrenic, however, is trickier, and the procedures employed often are so extreme (such as repetitive and unavoidable electroshock) as to make extrapolation of any findings to the human condition something of an act of faith. Nevertheless, such difficulties were shrugged aside, as potential drugs could now be tested on these depressed or anxious animals, and if they alleviated the symptoms could become candidates for testing on humans.

Over the decades, as neuropharmacologists and neurophysiologists discovered more and more neurotransmitters, each in turn became the molecule of the moment as a target for pharmaceutical intervention – acetylcholine, glutamate, dopamine, noradrenaline, gamma-amino-butyric acid, endorphins and, with increasing enthusiasm from the 1980s onwards, serotonin, the target for Eli Lilly's Prozac and GSK's Seroxat. Like chlorpromazine in the 1950s, the specific serotonin reuptake inhibitors (SSRIs) became the new wonder-drugs. For their most extreme advocates, like Peter Kramer, drugs such as Prozac would make a person 'better than well'.[14] Happiness at the pop of a pill offered the realisation of Brave New World. If only...

When the evidence began to pile up that the drugs aimed at single transmitter targets were not very effective, the theory changed – perhaps the problem lay in the balance between different transmitters? New cocktails of drugs were duly developed, such as serotonin-noradrenaline reuptake inhibitors (SNRIs). The trouble with both theories was that there is no evidence of disorders of neurotransmitter function in the brains of people diagnosed with psychiatric disorders; it is all an inference from the effect of the drugs, which certainly do have major effects on the transmitter systems. Living systems are resilient and respond to external insults by reorganising their cellular biochemistry. Thus brain cells respond to long-term drug administration by lasting changes in the efficacy of the neurotransmitters the drugs target. If not as extreme as in the case of chlorpromazine, the iatrogenic consequences can be severe.

There is evidence that the rise in psychiatric diagnoses may be in part a consequence of the long-term use of such drugs themselves. When conditions like depression and schizophrenia were first diagnosed in the last century, they were typically short episodes, which remitted and did not recur. Today they are seen as persistent and chronic conditions often beginning in very young children and recurring throughout life.[15] Critics such as Cardiff Professor of Psychological Medicine David Healy also argue that Pharma has been expanding its own market by working with the American Psychiatric Association to extend the number and variety of diagnostic conditions and then formulating drugs to match them. As its patent for one condition expires, the drug can be relabelled and patented for use in another. Healy points in particular to the rise in the diagnoses of Panic Disorder in the US after 9/11, which led to a relabeling of older drugs as anxiolytics.[16]

Certainly over the last decade evidence for malpractice by pharmaceutical companies in testing and marketing the drugs has mounted. Research papers ostensibly written or co-authored by senior academic scientists and clinicians turn out to have been ghost-written by the companies themselves. Leading academic psychiatrists have been shown to be covertly on the payroll of the Pharma companies, paid to be 'key opinion leaders' with their talks scripted for them by the companies. In response, the leading journals have insisted that authors declare their financial interests – but the journals themselves are tempted to publish favourable reports of drug trials, as the Pharma companies will then pay for thousands of reprints of the paper to be distributed to clinicians. Clinical trials conducted by the companies systematically emphasise the superiority of the drug under test to placebo or rival products. Negative data and adverse reactions are minimised or suppressed. Often the negative results have only come to light when NICE, the National Institute for Clinical Excellence, has requested all the data so as to conduct meta-analyses. The scandals, extensively documented by Healy, led the former editor of the *British Medical Journal*, Richard Smith, to call for all such trials to be publicly funded and openly published.[17] Unsurprisingly the pharmaceutical industry had never liked NICE, as its judgements on efficacy for health and value for money (intended to control which drugs the NHS could prescribe) impeded Pharma's sales efforts. The incoming Coalition government's response was to downgrade the Institute; the GP-based health consortia being assembled as part of the drive to part-privatise the NHS would be free to ignore NICE and choose what to prescribe. In a situation where the best and the cheapest drug are not one and the same, the market pressure on the new consortia is to maximise profit not quality of care.

The problem was that for all the advertising claims the evidence didn't stack up. The meta-analyses conducted by NICE showed that the SSRIs were no more effective than earlier generations of antidepressants. Reports of adverse reactions, even suicide among children, mounted.[18] Prozac featured in numerous court cases in the US until Eli Lilly settled with the claimants out of court for an alleged $50 million.[19] GSK was found to have suppressed evidence of the harmful effects of its SSRI, Paxil. An internal GSK memorandum stated that it 'would be unacceptable to include a statement that efficacy had not been demonstrated, as that would undermine the profile of paroxetine'.[20] In 2012, in what US government officials called the largest case of health care fraud in American history, GSK was fined $2.9 billion for mis-selling, bribing doctors and placing misleading articles in medical journals. Many patients found it difficult to come off the drug without suffering severe withdrawal symptoms – data also suppressed by the companies.

Even worse, the meta-analyses showed that none of the antidepressants, old or new, performed much better in the long term than placebos. What became increasingly evident was that the disordered-transmitter theory of mental illness was at best unproven. Like crutches, the drugs can be a support for people in pain and distress, alleviating symptoms but not addressing causes. And despite everything, the number of people being diagnosed with depression, bipolar disorder and other psychiatric conditions continues to rise.

NEUROGENETICS: WHAT WENT WRONG?

Biological psychiatrists began to look more critically at the DSM criteria. Suppose that a depression diagnosis refers not to a unitary condition but is a phenomenological state that could result from several distinct biochemical causes, each with a specific gene responsible for it. A drug that worked for one might not work for others, so the beneficial effect for a subset of the population would be submerged in a sea of negative results. But if the genes could be identified... This then was the pharmacogenetic hope with which Pharma and psychiatry approached first the HGP and then the biobanks. Once the genes responsible for these new disorders had been discovered, genetic engineering would enable them to be inserted into mice to make animal models for depression, schizophrenia or whatever. Rationally designed drugs tailored to individual genomes would be developed to replace those rapidly running out of patent.

If specific human gene mutations associated with neurological disorders can be identified, making the mouse models is relatively straightforward. A good example is Alzheimer's disease (AD). Mostly, AD is a disease of old age, but there is one rare form (affecting some 5 per cent of all those with the disease) that strikes in mid-life. This inherited form is associated with specific mutations in genes coding for the protein APP (amyloid precursor protein). Genetic engineering makes it possible to isolate the mutant genes and insert them into mice. 'Normal' mice don't suffer from anything analogous to AD, but the engineered Alzheimer mice and their offspring show specific difficulties in learning and remembering as they age compared with littermates. Potential anti-AD drugs could now be tested for their effects in preventing or at least limiting these cognitive declines in the mice.

Whether changes in a mouse's capacity to learn and remember a route through a maze can really be taken as an effective analogue for human AD is open to discussion – and anyhow the APP mutations in humans are only observed in rare early onset cases of the disease. For the vast majority of cases, the greatest risk factors are ageing, being a woman, and the gene variant *ApoE4*. Nonetheless, the Alzheimer mice and their well-defined learning tasks have become a standard drug-evaluation protocol for biotech companies and the pharmaceutical industry. The size of the market for an effective anti-Alzheimer drug is huge and growing. So the incentive for Pharma to invest heavily in AD (or more generally anti-neurodegeneration) research is high. Yet there are still only four drugs licensed for the disease, none of them very effective, and three of which have been available since the 1980s and '90s. Substances which in laboratory tests can improve a mouse's performance in learning to run around a maze do not seem to prevent memory loss in people with AD; still less do they work as smart drugs – cognitive enhancers – to improve a student's exam score (Ritalin for attention, Modafinil or coffee for wakefulness, are better bets with fewer adverse effects). Compounds that seemed promising after years of animal studies and the complex business of obtaining regulatory approval for human trials have had to be abandoned at huge expense as the negative results from the trials come in.

The problem, tough enough for neurological diseases, becomes much more acute with psychological disorders. Reports of major genes associated with depression or schizophrenia regularly appeared in the literature with much attendant publicity in the years prior to the HGP and the biobanks, only to be much less noisily withdrawn or forgotten as the failures to replicate multiplied. Once the biobanks had made gene-wide association studies (GWAS) possible, the reasons for the failure became obvious – many tens, perhaps hundreds, of genes, each with a small effect, could be associated with each diagnostic condition. Unfortunately, each new schizophrenia or depression GWAS based on a different population has thrown up a different array of genes. Discussing the problem of missing heritability, an article in *Nature* quotes Leonid Kruglyak, professor of evolutionary biology at Princeton, describing schizophrenia as 'a trait from which the genes have gone missing'.[21] This messy picture offers few if any clues as to new drugs, although Pharma still clings to the possibility as its last best hope.

THE MILITARY

Biomedicine and the pharmaceutical industry have not been the only ones interested in understanding the brain and mind. The military too has had a long-standing interest in agents that might enhance its own fighting capabilities while degrading those of the enemy, and has its own research budget for the purpose. Drugs that enhance attention, help keep troops awake, or diminish the horrors of battle have a long military history; the current favourites are Ritalin and Modafinil, plus the use of dope by soldiers, from the Americans in Vietnam onwards, to blot out the horrors of war (it is also an effective pain reliever, as MS sufferers have long known). Asked to predict future developments in neuroscience of military significance, the US National Research Council (NRC) gave high priority to the development of more effective cognitive enhancers.[22]

Against the enemy, there have been the nerve gases (German and British inventions from the 1940s and '50s, used with lethal effect by Saddam Hussein against the Kurds of Halabja in 1988) and botulinum toxin, the latter lethal at doses of three billionths of a gram. These chemical and biological weapons have been banned under successive international conventions ever since the 1925 Geneva Conventions formulated in revulsion at the horrors of the gas warfare of 1914–18. But the neurosciences promise something different: how about a drug that will incapacitate without killing? These could be regarded, like the tear gases, as outside the international bans – the argument used by the US to justify the massive use of the tear gas CS against National Liberation Front fighters in Vietnam in the 1960s. Also during that decade, as a result of their greatly expanded research budget, the US military hit on a drug that they claimed would disorientate and confuse an enemy. Codenamed BZ, it works a bit like LSD. A publicity movie made by the US military's Chemical Corps showed troops exposed to it collapsing in laughter, throwing their rifles down and ignoring military orders. Unfortunately, in practice, army chiefs seem to have been too cautious to employ it.

By the 1990s a whole new generation of incapacitants beckoned, euphemistically named calmatives for the outside world (and to the insiders spoken of as 'on the floor' or 'off the rocker' drugs). To avoid any suggestion that the research contravened international conventions, the funding has come from the civil budget, with the claim that the 'calmatives' were intended for crowd and riot control. International agreements only cover the use of chemicals in war, not for internal use to maintain order. The new incapacitants were quietly researched in many countries – notably the US and Israel – without attracting much attention until 2002, when one was used with disastrous effect and ensuing worldwide publicity by Russian special forces trying to rescue hostages trapped in a Moscow theatre. At least 130 hostages died when the opioid analogue fentanyl was pumped into the theatre's ventilation system. The episode proved, if proof were needed, that so-called 'non-lethal' weapons are in practice at best 'less-lethal'. On a much less dramatic scale, in separate incidents in the summer of 2011, two apparently fit young men in the north of England died after being sprayed with CS by police (one was also hit with a taser – a device increasingly used by police that fires a pair of thin wires with hooks at the end, which attach to the target's body or clothes and administer a painful and sometimes paralysing electric shock).

In the US, the 9/11 attack on the twin towers, and shortly afterwards the mailing of envelopes containing deadly anthrax spores to politicians, led to a panicked evocation of a new spectre – bioterrorism. A scarcely commented-on irony is that the anthrax spores turned out to have been 'militarised' and manufactured as part of the US's own defence research programme, and had apparently been stolen and mailed to the Congressmen by a disaffected former employee of the programme. Nonetheless, under the auspices of the Department of Homeland Security (a Volkish term characteristically aped by Blair in the UK, though without the attendant lavish funding) research budgets for all aspects of bio- and neuro-weaponry and bioprotection were dramatically ramped up. George W. Bush's bioterrorism budget for 2003 amounted to a staggering $11 billion. The manna was welcome to US university researchers, neurotech companies and industry alike, as more and more far-fetched projects were endorsed by a panicked government. An obscure reference in one of Steven's research papers was picked up by botulinum researchers at Edgewood Arsenal, the giant US chemical weapons research station. Memory researchers who had hitherto focused on discovering cognitive enhancers for AD now proposed to research memory-erasers that could treat the post-traumatic stress disorders of survivors from the twin towers and the increasing numbers of psychiatrically disturbed troops returning from Iraq and Afghanistan. Science-fiction indeed.

Neuropharmacology has been but a sub-theme within the military's interest in the neurosciences. From the 1960s onwards, almost the entire US university-based research programme in artificial intelligence was supported by grants from the US Defense Advance Research Projects Agency, DARPA. DARPA's interest was twofold. Might it be possible to enhance the capacities of combat troops (now rebranded as 'war-fighters') by tapping directly into their brains? Smart soldiers for the age of smart weaponry. A helmet incorporating electrodes able to read off the brain's electrical activity might serve the purpose. By the 1990s and more rapidly in the new millennium, the prospect of direct brain–computer interfaces began to seem practicable. By 2011 DARPA was investigating interfaces that could enhance the capacity to analyse intelligence information, improve motivation and accelerate learning by plugging into specific neural systems, as well as improve a war-fighter's ability to detect and identify threats rapidly and at a distance. In addition, the NRC proposed to supplement cognitive and sensory enhancements with robotic prostheses and orthotics designed to extend human physical performance.

The military's neuroelectronic ambitions go beyond enhancing its own fighting capacities into degrading those of the enemy. For decades DARPA has been interested in developing microwave radiation devices that beamed at opponents could disorient and pacify them, or, even better, read their intentions and modify their thoughts. (Back in the days of the Cold War it was called brainwashing, then mainly achieved through more classical techniques such as food and sleep deprivation and psychological assault.) Despite the hopes – and the suspicions of some US citizens that their government was indeed covertly controlling them via radiation waves – nothing practical resulted. By the millennium another prospect presented itself, exploiting the fact that communication between nerve cells in the brain is electrochemical. Where there is an electric current, there is a magnetic field at right angles to it – interfere with the magnetic field and the brain signals are disrupted. At the present stage of the technology, transcranial magnetic stimulation (TMS) requires the magnets to be placed directly around a person's head and focused on specific brain regions. In clinical trials such stimulation has been used to treat depression, obsessive-compulsive behaviours and Parkinson's disease. But what if it could be used at a distance to control thoughts or change behaviour? No wonder there has been a flurry of TMS-related DARPA contracts to US companies and universities.

The prospect may still be in the realm of science fantasy, but the mammoth military budget signed into law by President Obama is roomy enough to ensure that the surplus of contracts is not likely to diminish in the near future, and there will be plenty of neuroscientists anxious for research funding to bid for them. The efforts of an American neuroscientist, Curtis Bell, to attract signatures for a pledge from researchers not to use their neuroscientific research for military purposes met with a sadly small response in the political economy of the technosciences.

As ever, from skin grafts and antibiotics to brain–computer interfaces, war advances medical care. In the First World War only a tiny minority of psychiatrists used psychological approaches, and then only in relation to officers, in contrast to the brutal treatment meted out by the army and its doctors to the 80,000 shell-shocked British soldiers. By 1939–45, the term shell-shock had been banned, replaced by post-concussional syndrome and treated by occupational therapy and vocational training. Similar injuries among US and NATO forces in Iraq and Afghanistan, including those suffered by survivors of roadside bombs (IEDs), are recategorised as mild traumatic brain injury and treated with surgery, drugs and rehabilitation by physiotherapists. The associated condition of post-traumatic stress disorder is treated by counselling and psychotherapy. But with increasing knowledge of the biochemical and neurophysiological processes involved in memory formation, the US military is exploring the possibility of erasing the stressful memory by drugs or even TMS – a prospect satirised in Charlie Kauffman's 2004 film *Eternal Sunshine of the Spotless Mind*.

The increased survival rate for seriously injured soldiers airlifted back from the wars in Iraq and Afghanistan has added to the pressure to develop limb prostheses that could be operated directly via signals from the patient's EEG – detected via an array of electrodes woven into a 'hairnet'. Similarly, arrays of light-sensitive electrodes implanted in the visual areas of the brain could, more reliably than stem cell solutions, help the blind to see. One DARPA project seeks to restore memory loss in brain-injured soldiers by bypassing the damaged brain regions via computer inputs. If it works, civilian sufferers could benefit too.

MAINTAINING SOCIAL ORDER: PREDICTING CRIMINALITY

Maintaining social order is a central task for any state, and spies and informers have been among the oldest means of providing advance warning of any trouble to come. Being able to identify pre-emptively potential revolutionaries, criminals, murderers, robbers, even troublesome youth makes the task of preserving social order easier. Since Lombroso's late-nineteenth-century project of locating criminal propensity in skull shape, from sloping foreheads and ear shape to skull asymmetry, science has increasingly come to the aid of the state. Both the social and the life sciences have continued Lombroso's project. A long-term study by criminologists Eleanor and Sheldon Glueck, begun in the 1940s, compared delinquent and non-delinquent adolescent boys from poor urban areas in the US and, in a period that favoured social explanations, found that the best predictors of subsequent delinquency were home environment and parenting. One perhaps still relevant conclusion was that the less parents reasoned with their sons and the more they resorted to physical punishment, the greater the chance of subsequent delinquency.[23] As boys tended to fall into the kinds of delinquency and criminality troublesome to the state, while the delinquency of girls tended to be sexual and hence confined to the social order of the family, it was taken for granted that the study would exclude girls.

Today's preoccupation with criminal genes and hence criminal brains and minds goes back, as we pointed out earlier, to the early days of eugenics, with their dubious records of family histories and twin studies. Much of the more credible data in support of some genetic involvement has come from the comprehensive twin and adoption registers maintained in Sweden and Denmark. This showed that if the genetic father of an adopted child has a criminal record then the child has an increased chance of developing anti-social and/or criminal behaviour. However, this correlation only applied to small-time property crime, and not to violence.[24] The pattern puzzled the researchers. Could genes predispose a person to be a burglar but not a robber, a pickpocket but not a mugger? If so pre-emptive action could be put into effect, at least for children predisposed to be burglars and pickpockets. The question of which behaviours are identified as criminal is a highly social one – Prime Ministers who lead their countries into illegal wars, or bankers who speculate recklessly with other people's money, tend not to fall under such definitions.

Han Brunner and his colleagues focused on violence in a study of a Dutch family in which the men showed high levels of abnormal behaviour. The research reported eight men 'living in different parts of the country at different times across three generations' who between them had been involved in or displayed 'aggressive outbursts, arson, attempted rape and exhibitionism'.[25] The men exhibiting these very diverse behaviours ranging from the antisocial to the criminal all carried a mutation in the gene coding for the enzyme monoamine oxidase A (MAOA), involved in dopamine and serotonin neurotransmission. MAOA became widely reported as the 'violence gene'. Brunner's original work has not been replicated, although in 2010 a *Nature* paper reported a mutant form of a gene associated with serotonin metabolism in a sample of violent offenders in Finland. It was described as a gene for impulsivity which predisposed the offenders to violent behaviour, although the ninety-six offenders studied also showed a variety of other forms of psychiatric disorder, and their violent crimes were almost all committed while drunk.[26]

In 2002, Avshalom Caspi and his colleagues came to a more complex conclusion, which seemed to bridge social and biological explanations. They studied a thousand twenty-six-year-old men, a New Zealand cohort which had been assessed from the age of three onwards. Some of these reported having been abused when children, and as adults a proportion were violent or antisocial – a judgement based on criminal convictions or personality assessments. The research found that men who had been abused as children were more likely to be violent or antisocial as adults. Caspi looked for associations between criminal and antisocial behaviour and MAOA activity. Men who had been abused in childhood and who carried the gene variant producing high levels of MAOA were less likely to be criminal or antisocial as adults than those with the gene variant producing low levels of activity.[27] The researchers went on to suggest that different forms of a gene associated with serotonin neurotransmission moderated the relationship between life stress and depression.[28] Caspi's work too has been challenged; several attempts to replicate it found that the MAOA variant made no difference except when people had been subject to childhood abuse. It was the abuse, not the gene, which predicted outcomes.

Whatever the causes, the question that concerned the experts was whether such antisocial behaviour could be predicted and prevented.

Untreated ADHD, the US Department of Justice claimed, is a predictor for subsequent delinquency. Increasingly in the UK and the rest of Europe as in the US, and in defiance of NICE guidelines for assessment, children are being diagnosed with the condition on the basis of parents and schoolteachers' reports. A prominent British educationalist describes the symptoms of ADHD thus:

> children with AD/HD ... disturb their parents and teachers because their classroom achievement is erratic ... a source of exasperation to the ... teacher. On occasions the child may show high levels of performance, a ready wit and an imagination of a high order but with erratic performance. This is the pupil who never seems to be in his or her seat, who is constantly bothering classmates, and can be relied upon for little other than being generally off task. All categories can be frustrating to teach because of their apparent unpredictability; their failure to conform to expectations, and their tendency not to learn from their mistakes.[29]

A mischievous child? An intelligent child bored by school? A child suffering from poor parenting? The child's behaviour is no longer perceived as part of a relationship – between him (they are mainly boys) and his school or parents – but as rooted in his brain. Whatever the cause, the child is a nuisance and must be medicated to control his behaviour.[30] The drug of choice is methylphenidate, a substance related to amphetamine and marketed as Ritalin, used to enhance concentration and attention. Ritalin has also become the poster-drug for a new generation of cognitive enhancers. The FDA warned that the prescribed drug was being traded in school playgrounds, although it can be readily bought off the web without prescription. In response to a reader survey by *Nature* in 2010, several hundred US academics and students reported using Ritalin and Modafinil to enhance their performance. Pharma has done well out of this culture of enhancement; over a period of fifteen years the legal production of Ritalin has increased seventeenfold and amphetamine thirtyfold.

Whether giving children Ritalin prevents subsequent delinquency or not, neuroessentialism maintains that behavioural outcomes are embedded in brain processes, and it follows that, even if genes cannot predict violence, perhaps brain scans might. Particular attention has been paid to criminals diagnosed as psychopaths – the diagnostic criteria for which include being glib, superficially charming, liars, callous and manipulative, failing to accept responsibility, impulsive and lacking remorse (criteria against which some successful entrepreneurs, media moguls and politicians might also be judged). As it is not possible to detect and study such people before they have actually committed any crime, the research has been carried out in prisons. US researchers compared MRI scans of twenty violent psychopaths in prison with those of other prisoners and claimed to find in the psychopaths abnormalities in the connections between the ventromedial prefrontal cortex and amygdala, two brain regions associated with empathy and emotion.[31] By contrast a study by former UK Home Office forensic psychiatrist Adrian Raine, now at the University of Pennsylvania, claimed to find differences in two other brain regions, the hippocampus and corpus callosum, of his criminal psychopaths.[32] Not only do the findings conflict but also they are post-hoc; the people studied have histories of crime, imprisonment, drugs and alcoholism, all of which may have affected brain structures. As such, the predictive value of the brain scans is dubious – although David Blunkett, when Home Secretary in the Blair government, mused aloud as to whether it might be possible to pre-emptively diagnose and incarcerate psychopaths before they had committed any crime (a wish more indicative of his inveterate authoritarianism than of his understanding of the technical possibilities or his commitment to the fundamental right of innocent until proven guilty).

While MRI scans provide static images of brain structures, fMRI is dynamic, and it was not long before it too began to be pressed into service in pursuit of juridical aims. A company called Brain Fingerprinting responded to Bush's post-9/11 'War on Terror' by offering to use an older technology to detect an electrical signal from the brain, called the P300 wave, to determine whether a suspect under interrogation is a terrorist or a spy. According to its company website, still extant in 2012:

> There is a new technology that for the first time, allows us to measure scientifically if specific information is stored in a person's brain. Brain Fingerprinting technology can determine the presence or absence of specific information, such as terrorist training and associations. This exciting new technology can help address the following critical elements in the fight against terrorism. [It can:] Aid in determining who has participated in terrorist acts, directly or indirectly. Aid in identifying trained terrorists with the potential to commit future terrorist acts, even if they are in a 'sleeper' cell and have not been active for years. Help to identify people who have knowledge or training in banking, finance or communications and who are associated with terrorist teams and acts. Help to determine if an individual is in a leadership role within a terrorist organization.

It isn't clear whether the CIA or FBI took up the invitation, and one might speculate about how Professor Sir Anthony Blunt, art historian and keeper of the Queen's pictures, and also one of the five long-serving Cambridge Soviet spies, might have stood up under the technology. Brain Fingerprinting's CEO, Larry Farwell, has claimed that the technique has been used in court cases to help determine a person's innocence. Ruled inadmissible in the US, EEG technology was used in at least one court case in Mumbai, India. As brain fingerprinting disappeared into the dustbin of discarded snake oils, an alternative emerged, using not EEG but MRI. NoLieMRI, a San Diego-based company established in 2006, proposed to use the scanning technique for lie detection. The response of the neuroscience and neuroethics communities has, however, been overwhelmingly hostile. The claims for the technology are premature, its reliability is low, and false positives and negatives are all too likely, said an Editorial in *Nature Neuroscience* in 2008.[33] Nonetheless, in 2010 the NRC noted the strong interest of the intelligence community in improving their ability to detect deception, and gave high priority to research aimed at finding 'neurophysiological indicators of psychological states and intentions'.[34] The trouble is that in a fearful society, with the dominant claims and imaginaries of an essentialist neuroscience ringing in the ears of the lawmakers, snake oil can look very attractive.

NEURO-ENTHUSIASMS

For lawyers the issues raised by the neurosciences are less about new technologies than the much broader question of where a reduction of mind to brain leaves the issue of legal responsibility. The central legal concept of *mens rea* is challenged by the claim that 'my brain made me do it'. Young children are exempt from legal responsibility, and so too are others judged legally incompetent by virtue of brain damage or – dating back to the M'Naghten Rules of the late mid-nineteenth century – insanity. Some developmental neuroscientists argue that adolescents' brains are still so immature as to make them risk-prone and therefore not fully responsible for their actions. As judge Stephen Sedley has pointed out, 'the law on human responsibility for acts which harm others is a set of historic and moral compromises'.[35] An accused's pleading that they were acting under the influence of recreational drugs or alcohol is inadmissible in UK courts. Will the situation be any different if a brain biomarker (such as an altered allele for an enzyme of neurotransmitter metabolism or an altered connection between different brain regions) could be found that predisposed a person to criminal activity? This defence has been tried – and rejected – in the US, and it seems likely that it is an argument that the courts will continue to resist.[36]

Such scepticism has done little to dampen the neuro-enthusiasm elsewhere. Art, literary and music critics – and indeed artists, novelists and musicians – offer to decode the popularity of artworks, music and novels in terms of how they resonate with particular brain structures, themselves the product of deep evolutionary history. This claim goes back to sociobiologist E.O. Wilson,[37] who suggested that we are evolutionarily predisposed to respond to landscapes containing grass, trees and water – mercilessly deconstructed by the architectural critic Charles Jencks, with his characterisation of Wilson's favoured genre as 'Battersea Road art'.[38] But perhaps the most striking intervention of neuroscience into day-to-day life has come from yet a further hyphenated field – neuro-education.

For neuro-educationalists, evidence from animal learning is increasingly being taken as providing clues as to how best to educate children. 'Brain gyms', along with dubious nutritional supplements claimed to improve IQ, have now become big business, pushed by heavy direct-to-consumer television advertising and endorsement by celebrities free from any knowledge of the neuroscience.

In laboratory trials, creatures as diverse as fruit flies and mice can be trained to recognise and avoid particular odours, or to change their behaviour when warned of an impending electric shock. It normally takes many trials for the flies or mice to behave reliably in the way the experimenter has taught them, but they learn faster when set a group of trials in quick succession followed by a rest period and then another group (massed training) than if the same number of trials is spread out more evenly (spaced training). As written up in the popular press, the finding persuaded a headmaster in the north of England to re-jig the school timetables. The children were now to be taught in ten-minute bursts, followed by a rest and then another teaching session. Applying evidence gained from studying the training of animals and insects to the education of children sets aside the same ethical and technical problems of translation as those involved in moving from the identification of a drug which works in animals to the point where it is proven both safe and effective for human use.

A more fundamental problem is the increasing conflation of training and education. Although the two can overlap in terms of the acquisition of specific skills and specific types of behaviour, education still carries more of the sense that only the untranslated and untranslatable German concept of *Bildung* offers. This refers to both the dimension of teaching and that of learning – not just knowledge and skills, but also values, ethos, personality, authenticity and humanity. Here, educational fulfilment is achieved through practical activity that promotes the development of one's own individual talents and abilities and which in turn leads to the development of one's society. In this way, *Bildung* does not simply accept the socio-political status quo, but includes the ability to engage in a critique of one's society, and ultimately to challenge it to realise its own highest ideals. The more pragmatic Anglophone culture would eschew such high-flown language, yet most educationalists and parents would say they wanted their children to develop these precious qualities – an ideal which is hardly to be achieved by contextualising them as if they were fruit flies or laboratory mice.

Nonetheless, such education neuromyths proliferate. Teachers report receiving up to seventy mailshots a year promoting a variety of neurononsense – critical periods in development, children as left-brain/right-brain learners or as visual versus verbal. The snake-oil entrepreneurs are in there selling hard to teachers who are without the protection provided by clinical trials and NICE for doctors' prescribing practices. Against this, a series of reports over the past few years – produced by the UK Economic and Social Research Council, the Organisation for Economic Cooperation and Development and most recently the Royal Society – have brought neuroscientists and educationalists together to counter the myths, resist the snake oil merchants and, most positively, envisage neuroscientists and educationalists as equal collaborators in the framing of research.

The weakness of these reports is that they avoid speaking about the elephants in the room – those academic neuroscientists who energetically proselytise, offering, like the geneticists and the stem cell biologists before them, their Promethean promises. However, where the geneticists and the stem cell researchers could speak with a unified voice, the umbrella of the neurosciences includes totally different approaches, from the invasive research of neurochemists and neurophysiologists to the non-invasive fMRI researchers and the black boxes of the psychologists. Neuroscientists don't all speak with one voice, but rather in a babel of contradictory languages and claims within a multiplicity of lobbies in an intensely marketized economy where the hope of profit trumps evidence.

9

Promethean Promises: Who Benefits?

In 1818 the brilliance of Mary Shelley's Gothic imagination transformed the Titan Prometheus into the scientist Victor Frankenstein. Since then, the imaginaries of the late twentieth and early twenty-first centuries have fostered another transformation: the Titans have become Arthurian knights on a quest for the Holy Grail. The molecular geneticists were the first such self-appointed knights, their Grail the Human Genome. As Harvard's Walter Gilbert put it at the launch of the HGP in 1991, 'the search for the Holy Grail of who we are has now reached its culminating phase'. For Genentech's founder, Herbert Boyer, the sequence was 'the Holy Grail of genetics'. Other molecular biologists followed suit and it was not long before the metaphor was adopted by all, recruited for every enquiry from plant genetics to susceptibility genes for breast cancer.

And so it moved yet further, from genes to cells and then to brains. Stem cells, for John Sinden of ReNeuron, would become 'the Holy Grail of determining if embryo cloning is possible in humans'.[1] For Peter Braude, at King's College London, the human embryonic stem cells his team has produced are 'really the Holy Grail for everybody in regenerative medicine'.[2] And for Leo Furcht from the Minnesota Medical School, they are 'the silver bullet, the Holy Grail of medicine'.[3]

As usual the neuroscientists have not been far behind. Their Grails range from the sublime to the trivial. In a video, the leading neurologist and author V.S. Ramachandran explains that research on 'what we call self-awareness [is] the Holy Grail of neuroscience'. Henry Markham of the IBM funded Blue Brain project and computational neuroscientist Giorgio Ascoli of Mason University see the construction of a 'virtual brain' as neuroscience's Holy Grail.[4] Other Grails are more practical: Daniel Chain, CEO of the neurotech company Intellect Neuroscience, sees 'a prophylactic vaccine [as the] the Holy Grail of Alzheimer's disease management'.[5] And some are absurd: Stephen Liberles, a Harvard neuroscientist, having discovered a novel molecule in carnivore urine that triggers an avoidance response in mice, described his finding as a step on the route 'from chemicals to receptors to neural circuits to behaviors is a Holy Grail of neuroscience (sic)'.[6]

In the quest for the Holy Grail, the chalice to hold the blood of Christ could only be won by the purest from among the Knights of the Round Table – Sir Galahad. But Galahad's purity is a far cry from today's treasure-hunting scientists, armed with their share options and company directorships secure beneath their breastplates.

The last few decades have certainly been exciting times for biologists. The life sciences, once a village of cottage industries, have been transformed into huge technoscientific enterprises, lavishly funded from many sources; a transformation only made possible through their fusion with informatics. New approaches and understandings of fundamental scientific questions about the nature of inheritance, the processes of development and the workings of the brain that have lain unanswered since the very birth of science have energised once-dormant fields of research. Data has cascaded out of the labs and into the production systems of both small-scale biotech and giant pharmaceutical companies. New breeds of scientific entrepreneurs have emerged from the universities and stormed into the corporate world. The pronouncements of leading biotechnoscientists are listened to with the respect previously given only to the most eminent of the nuclear physicists; today they advise governments and industry, and give well-received lectures to the world leaders at Davos.

Their language has entered culture. Hardly a day passes without metaphors of DNA, hard-wiring, Darwinian natural selection and evolution being invoked by the media or politicians, to say nothing of the rest of the chattering classes. This deference to the determining authority of the life sciences spreads ever further: philosophy, art, ethics, sociology, politics and law all feel the need to position themselves for or against its claims. Does identity lie in the genes, the neurons or the Pleistocene past? Has the neoliberal self arrived, demanding that the Promethean promises of regenerative medicine deliver?

Perhaps the bluntest assessment of the gene Grail comes from one of the key figures in the HGP. Eric Lander is quoted as saying, in a speech delivered at the White House: 'We've called the human genome the blueprint, the Holy Grail, all sorts of things. It's a parts list. If I gave you the parts list for the Boeing 777, and it has 100,000 parts, I don't think you could screw it together, and you certainly wouldn't understand why it flew.'[7]

The quest for the stem cell Grail is still ongoing, now directed towards adult rather than embryonic stem cells. As for the Knights-errant of neuroscience in their quest for consciousness and the self, our view is that they are galloping off to nowhere. Minds cannot exist without brains, but are not reducible to them.

TOO FEW GENES, AND TOO MANY

Unfortunately, genes could never be the Holy Grail, but offer a far more human story. At the initiation of the HGP, no one knew how many genes were embedded within the three billion A's, C's, G's and T's of the human genome. The favoured guess was about a hundred thousand, and as the race got under way, the sequencers ran a sweepstake with the prize going to the one who came closest to the right answer. Once decoded, they were confident the sequence would reveal how the genes it contained could generate the 100 trillion cells in every human body, with each of the 7 billion humans on the planet having a unique analogue readout from a digital DNA string. The race to complete the sequence – with its conflict between public and private, between the sequence as an open resource on which Pharma could draw to come up with its patents, and the sequence as a closed book, access to which would be controlled from the start by private gatekeepers – became the archetype of the contradictory forces within the new political economy of science.

More than just a scientific triumph, the molecular geneticists argued, decoding the genome would open the door to a cornucopia of benefits in terms of health and wealth. Personal genomes indicating health risks would be transcribed onto CDs, those glittering discs held up by the molecular biologists to spellbound audiences as they sought to secure support for the first macro project of the life sciences. Well before even the first draft of the genome was published in 2001, plans for the biobanks were under way, spearheaded by Iceland's deCode. With the costs of whole-genome sequencing still running into billions, the biobanks proposed to look more modestly for SNPs – variants in genetic sequence shared by people with a particular common disease – with the intention of opening the door to both predictive and personalised medicine through pharmacogenomics.

Despite the hope that enough information could be collected about individuals to enable the biobankers to identify environmental as well as genetic factors for a disease, so far, the emphasis has been on decoding genes – on the 'hard data' of DNA analysis rather than the 'soft' data characteristic of complex physical, social and cultural environments. Furthermore, the biobankers have only been able to sample a specific subset of the richness of human genetic variation – overwhelmingly they have thus far studied only the well-educated genes of European descent.

In the UK, the fact that UK Biobank extended its invitation to general practitioners' patients without weighting for ethnic minorities means that the diversity of bio-geographical ancestry, to say nothing of class, is underrepresented. The MRC is unconstrained, and appropriate representation has been left to the good conscience of the researchers. US science policy, by contrast, politically confronted by the failure of biomedicine to study women and minorities, requires that both are appropriately represented in Federally funded research.

As the results of the HGP began to come in, Michael Dexter,

the CEO of Wellcome, claimed that the completion of the project was more important than putting a man on the moon, on a par with inventing the wheel. In fact, the results were something of a theoretical embarrassment to genocentrism. Humans, supposedly the pinnacle of evolution, with the most complex of brains, turned out to possess only some 20,000 genes – about as many as a fruit fly. The molecular biologists who had confidently predicted that all human life could be read off from the linear string of DNA went rather quiet.

In response, Lander and Craig Venter went public, upgrading their Grail quests from individual genes to genomes, proteomes, and the rich developmental interactions of epistasis and epigenetics. The reactions from the rest of the natural science community varied from the flat, if slightly surprised, reporting of the major journals to brushing the problem away with a joke – it was 'amusing' that we shared half our genes with a banana. It was left to the anthropologist Jonathan Marks to reflect more seriously on *What it means to be 98% chimpanzee*.[8]

For the biobankers things only got worse. Not only were there too few genes, but, rather than only a handful being involved in any disease condition, it turned out that there were many. Even as UK Biobank finally reached its recruiting target and opened for business in 2011, the results of the initial gene-wide association studies (GWAS) were beginning to show that disease associations were much more complex than the geneticists had predicted. Multi-authored papers poured out from the international consortia, published in leading journals. The genetics was fine (though the environment had vanished from the account), but in terms of health or wealth deliverables, the results were minimal.

DeCode's high-profile papers did not save it from bankruptcy, while the biobankers sought a technical solution by increasing their overall population size through international networks. Sceptical geneticists spoke of the problem of 'missing heritability'; others, notably the editor of the journal *Genetics in Medicine*, James Evans, and the health psychologist Therese Marteau, argued that the genetic bubble had burst.[9] Could it be that the vast edifice of funding and skill that the genome project and the biobanks had built was based on shaky theoretical foundations?

Even though Francis Collins, Director of NIH and a central figure in the successful sequencing of the human genome, tempered his initial euphoria, he remained optimistic, rather like Dickens's Mr Micawber, that something would turn up.[10] Having claimed that within a decade of completing the genome the new knowledge would transform clinical care, when it didn't, he shifted his argument. Now he concedes that although the promises of new technologies tended to be oversold by the early enthusiasts as delivering too much too soon, in the long run they will nonetheless exceed the best hopes of the pioneers.

That nothing turned out to be quite so simple came as no surprise to those unhappy with genocentrism. What the HGP and the biobanks have revealed is a degree of biological complexity long suspected by less reductionist biologists but to which the sequencers and their backers, in their enthusiastic and single-minded pursuit, had turned a blind eye. Lander was right: the completion of the sequencing of the human genome has produced a list of parts with no instruction book on how to put them together. While the theorists of genetic determinism were seemingly unmoved by the paucity of genes, the Landers and the Venters simply moved on.

The dramatic decline in the cost of genomic sequencing has meant that it will soon be possible for anyone with the odd $1,000 to spare to have their entire personal genome inscribed on a CD – or perhaps these days downloaded onto a smart phone. This is a major step beyond the dubious gene markers offered by 23andMe and its competitors. However, this torrent of bioinformation is of little use to the patient or their clinician unless Pharma can design a drug tailored to treat a patient's specific genetic variant. But as the results of the GWAS for complex diseases are proving, this is unlikely.

This genetic complexity generates risk assessment with little predictive value, a phenomenon troubling to medicine and social theorising alike. The loss of predictive value brings difficulties to the thesis of biological citizenship proposed by sociologists Carlos Novas and Nikolas Rose, who wrote:

> The responsibility for the self now implicates both 'corporeal' and 'genetic' responsibility: one has long been responsible for the health and illness of the body, but now one must also know and manage the implications of one's own genome. The responsibility for the self to manage its present in the light of a knowledge of its own future can be termed 'genetic prudence'. Such a prudential norm introduces new distinctions between good and bad subjects of ethical choice and biological susceptibility. This contemporary biological citizenship operates within what we term a 'political economy of hope'. Biology is no longer blind destiny, or even foreseen but implacable fate. It is knowable, mutable, improvable, eminently manipulable. Of course, the other side of hope is undoubtedly anxiety, fear, even dread at what one's biological future, or that of those one cares for, might hold.[11]

As the GWAS results began to appear, the possibility of genetics providing individuals with their personal genetic risk receded. For most of us, is there any specific self-disciplining action that the individual can take to influence their future? Are there any new recommendations over and above adopting the routine advice of twenty-first-century health promotion?

Genetic certainty is only for those at risk of dominant single-gene diseases; and here the messages are all too clear. Take Huntington's disease, where Novas analysed the online discussion of a group of at risk people. The gene for this early onset neurological degenerative disease was identified thirty years ago; since then, despite all the hype, there have been no therapeutic deliverables, no gene therapy, no drugs – only testing. That so many at risk reject testing speaks of a decision not to know when there is no effective action that anyone can take to change their genetic future. Better stoical uncertainty.

The situation is different with another single-gene disorder, Familial Hypercholesterolaemia (FH), which elevates cholesterol levels and hence increases risk of premature morbidity and mortality from heart disease. Here there are effective drugs which, combined with life-style changes, can and do reduce risk. Hilary, having FH, was both a member of the FH Association and, with their knowledge and agreement, researching the group. The FHA provided mutual support, but its members almost never spoke of genetics. Instead they were more likely to swap life-style tips, chiefly about dietary management, as well as stories of radically different therapies designed to reduce their dangerous cholesterol levels. For some in the group, their shared genetic risk had become their polis, their civic responsibility limited to personal risk reduction. Perhaps their risk-preoccupied discussion was context dependent, but even in the setting of a small group, meeting around the dining table of the vicar's wife (her brother had FH), little else was talked about.

Hilary, despite her well-founded anxiety, found herself irritated by this restricted polis; for her there was more to life and citizenship than FH. Once the research study was over she left the group, but, just as Novas and Rose proposed, she had become an active well-disciplined genetic citizen (mostly) restricted to just the FH part of her life. Active biological citizenship requires well-founded hope, a hope that the complexities and predictive weakness of GWAS that now dominate the genetic explanation of common diseases cannot provide.

The mantra of personalised medicine thus seems to fit better with the ideology of individualism than with the realisable prospects of genomics. Whatever the specific gene variant(s) involved, the most practical strategy for the clinician is still to use one of the existing broad spectrum drugs. Although several hundred different mutations have been identified in the single gene associated with FH, for most people the statins are an effective treatment. Against the imaginaries of personalised medicine, the one-size statins pretty well fit all. As the statins come out of patent and their price comes down, what is good news for patients and health-care providers is bad news for Pharma's bottom line.

It is difficult not to reflect here on a study part-funded by the British Heart Foundation which recommended that statins be prescribed to 20 million of the UK population.[12] This proposal looks set to medicalise even more of the population, increasing Big Pharma's market. Such a policy is a far cry from the WHO's 1980s European Health Promotion programme, which sought to reduce the very high levels of heart disease and premature death through encouraging people to exercise and reduce the saturated fat in their diet. Finland in particular achieved remarkable improvements. By contrast, obesity, a condition associated with heart disease, diabetes and premature death, is at record levels in Britain. It is unlikely that genetics research or even a pill will solve this complex problem.

SO WHO DOES BENEFIT?

Big Pharma's profits have continued to rise. In 2010 the top twelve US pharmaceutical companies in the Fortune 500 had a combined revenue of $250 billion – an increase of 5 per cent over the preceding three years, while over the same period their profits had risen by an average of 25 per cent; only the largest of all, Johnson & Johnson, posted a small decline. In the absence of new blockbuster drugs and with the expiry of present patents, Pharma knows it has a problem. Its response, given that innovation is its life-blood, is contradictory. A recent survey of the twelve largest global Pharma companies revealed a drop in R&D expenditure and a closure of research facilities as more and more drugs failed in the advanced stage of their clinical trials. In 2010 only twenty-three new drugs had reached the advanced trial stage; by 2011 the global total had dropped to eighteen. Meantime the cost of bringing a new drug successfully to market has risen from some $800 million to $1.05 billion.

To maintain sales, Pharma turns to its established practice of finding new targets for pre-existing compounds. One way to do so is to create a new disease category for which the drug is said to be effective, as in the case of 'Panic Disorder' after 9/11.[13] An alternative is to claim the drug is effective only for a specific group, as with BilDil, a combination drug that failed its mixed-population clinical trials but has been approved by the FDA for treating congestive heart failure in a specific group: African Americans – whose African ancestry is associated with an elevated risk of heart disease. BilDil was promoted as a drug embodying the NIH's commitment to the policy that biomedicine should serve minorities better. Despite this, BilDil ran into controversy, its critics arguing that racialised prescribing on the basis of self-reported racial identity was unsafe.[14]

As the previous chapter documents, today Big Pharma predicts that the prospects of finding new, effective psychotropics are bleak. The hype and hope at the launch of the HGP that genomics would deliver cures for schizophrenia and other severe mental illnesses has faded, and several of the major pharmaceutical companies are closing down this line of research, even as the world pandemic of unhappiness – depression and anxiety – increases year on year. While the psychotropics can help a patient through an acute psychotic episode or to manage depression, their long-term use, producing as it does a vicious cycle of tolerance and almost irreversible brain changes, is problematic. As the critics of biological psychiatry argue, *Your Drug May Be Your Problem*,[15] and this may be even more true when taking a cocktail of drugs. Even among those previously committed to biological psychiatry, voices have begun to be heard arguing that the whole approach of treating mental disorder and distress as if it could be cured or alleviated by manipulating brain chemistry is misguided – another fractured paradigm.

Big Pharma's failure to innovate has not been reflected in the remuneration of senior executives. In 2011 the UK High Pay Commission reported that executive pay across the entire industrial and commercial spectrum had increased by 49 per cent in the preceding year, while that of the average employee rose by only 2.7 per cent. CEOs in the FTSE 100 received an average £3.8 million. (By 2012 it had risen to £4.8 million.) For comparison, the national average salary in 2010 was £25,900. For directors, the earnings element in the total package is less than the add-ons: the taxable benefits, pensions and bonuses; at 40 per cent of the total take, bonuses alone are greater than basic earnings.

The CEOs of Pharma companies in the FTSE 100 rank second only to those in the oil and gas industries. Despite the economic crisis, their remuneration has continued to rise – astronomically in some cases. In March 2012 there was a public row over the £6.7 million awarded to GSK's CEO Andrew Witty. The amount had been doubled to 'keep GSK competitive' in the global CEO market. Even then the sum was deemed inadequate; despite the evidence piling up in the US during this period against GSK's fraudulent mis-selling (so far the UK has taken no action), Witty negotiated an increase to £10.4 million on the grounds that his remuneration still left him in the bottom quartile of Big Pharma CEOs. Indeed, he is poorly rewarded by US standards; Johnson & Johnson's CEO, former marketing director William Weldon, saw his total remuneration increase to $27 million in 2010. The efficacy of the incentive was dubious as J&J's profits fell a year later.

Such obscene sums of money may well have rather little to do with running an effective company and rather more with the composition of the remuneration committees, greed and the symbolism of high pay as a measure of masculinity. Women are in short supply at the directorial level and the best-paid woman gets slightly less than the worst paid man. None of Big Pharma's CEOs have a background in the biomedical sciences; only 43 per cent of UK boards include a Research and Development Director, and even where there is one, their remuneration is well below that of the CEO or even the Chief Finance Officer.

Although CEOs bonuses seem to go ever upward, there are signs that the so-called 'Shareholder spring' is having an effect. Faced with a catastrophic decline in profits, the CEO of AstraZeneca, marketing man David Brennan, resigned before he was pushed, to 'spend more time with his family'. His departure was no doubt eased by his having received more than £9 million in pay and perks in 2011 followed by a £5 million retirement package.[16]

Pharma's problems are echoed in other biotechnosciences. Although Amgen, the world's largest biotech company, with top-selling drugs for cancer and anaemia, was set for an income of $5 billion in 2012, many others have never turned a profit. Among the leading stem cell companies, Geron, which recently closed down its much publicised stem cell trial, is capitalised at $630 million, Osiris Therapeutics at $474 million, and StemCells Inc at $174million. None yet has a product on the market or has registered a profit.

Geron hired a new CEO for its stem cell venture, biotech executive John Scarlett. His remuneration is not declared but his previous position included $7 million of stock options, so he is unlikely to have taken a cut. By contrast the companies that are currently on target to make a profit are those that manufacture the equipment and synthesise the chemicals needed by the stem cell companies. The same is largely true for neurotech. The companies that generate most of its $4 billion turnover produce the equipment that university and public sector research require. But the health benefits from this research have yet to materialise.

The hybridity of the biotechnosciences has meant an increasing permeation of the values of industry into academia. Over recent decades, the heads of the UK's Research Councils, once modestly called Secretaries, have mutated into CEOs. The Councils' chairs and many board members are drawn from industry and finance. The Chair of the Economic and Social Research Council is an investment banker. The Medical Research Council is chaired by Sir John Chisholm, who has no background in medicine or the life sciences but was the civil servant in charge of selling the government's defence research labs to a private company, QinetiQ, of which he then became Chairman. The MRC's governing body includes – as *Guardian* columnist George Monbiot has pointed out – 'executives or directors from Pfizer, Kardia Therapeutics and Microgen Ltd, but no one who makes their primary living working for a medical charity or any other public interest group'.[17]

Just to ram the point home, the high point of the Euroscience Open Forum meeting in Dublin in 2012 was a session introduced by the Irish Minister for Research and Innovation entitled 'Science-2-Business'. The keynote lecture was given by the biochemist Pearse Lyons, founder and President of the biotech company Alltech, and billed as being about 'how he took $10,000 and transformed it into a Billion-Dollar Business operating in over 120 countries, over 30 years, without recourse to venture capitalists or outside investors, by always maintaining primacy in science'. Or, as science journalist Tom Abate, writing in the *San Francisco Chronicle*, put it: 'When we spliced the profit gene into academic culture, we created a new organism – the recombinant university. We reprogrammed the incentives that guide science. The rule in academe used to be "publish or perish." Now bioscientists have an alternative – "patent and profit."'[18]

WHO PAYS FOR BIOLOGICAL DETERMINISM?

Throughout the last century genetics and eugenics have been conjoint twins, their histories intertwined. The naive genetic determinism of the early part of the century, which saw moral degeneracy, feeblemindedness and criminality as located in single genes, weakened in the late 1930s and by the mid '50s had been abandoned by mainstream genetics. But even as academic geneticists sought a clean surgical separation from their now deplored twin, the medically led eugenic practice of compulsory sterilisation continued, in the US mainly directed against African Americans, while in Europe the Nordic countries sterilised women 'unfit to mother' and the British and Dutch continued their policy of sexual segregation and incarceration.

With E.O. Wilson's *Sociobiology* genetic determinism returned, not as eugenics but as an ideological justification of the existing social order with all its injustices. The ensuing furore between the self-confessed conservative Wilson and his radical and feminist critics made it impossible for his claims to be seen as other than controversial.

Evolutionary psychology, understanding itself as a scientific advance over sociobiology, came to the rescue of biological determinism, arguing that human nature had been 'fixed' in the Pleistocene, and that there hadn't been enough evolutionary time to change that nature subsequently. The important modification that evolutionary psychology made was to insist on the unity of humanity and abandon racialised difference. Nonetheless, despite the evolutionary psychologists' claim to being Darwinists their determinist stance directly conflicts with the radical indeterminacy of Darwin's theory of evolution.

By the millennium, as neuro-imaging flourished and moved beyond its early primarily medical-diagnostic uses, its findings were increasingly summoned into the debate as evolutionary psychologists and cognitive neuroscientists combined forces within the new social neuroscience. By assigning mental attributes to brain locations, they began to generate an internal phrenology in which our self-conscious thoughts, feelings and decisions are no more than the epiphenomenal products of brain processes that had themselves been selected. Thus they returned to some of the earliest preoccupations of Darwinian evolutionary theory: the origins of altruism, moral judgement, art, music, racism, love and consciousness. These were all, they proposed, the products of natural and sexual selection some 200,000 years ago. The significance of history – in its double sense as both the passage of time and as a discipline exploring and interpreting human cultures and societies – is erased. It is as if evolutionary psychology claims to speak not only the truth but the whole of truth.

Some of this internal phrenology, such as Francis Crick's location of free will in the anterior cingulate sulcus or Semir Zekir's siting of romantic love in the insula and putamen, is gender neutral. Elsewhere in the neurosciences scientific sexism is making a disquieting cultural comeback with a renewed focus in the interactions of hormones, brains and gender. Among the hormones discovered early in the last century were two seen as sexually specific: testosterone and oestrogen. These names were less than subtle, signalling as they did that there was little possibility of them doing anything except determine maleness and femaleness – for men and women, masculinity and femininity – now known to be far from the case. In the 1970s this was inflated with the biopolitical claim that men's testosterone drives them to pursue success while women are at the mercy of oestrogen. This claim immediately stoked criticism from both biologists and deconstructionists active in the women's movement, who questioned the determinists' knowledge of the complexity of current hormonal theory. Biologist Gail Vines denounced them as traducing the interactions in the body by reducing them to a crude either/or,[19] while deconstructionist Nelly Oudshoorn[20] pointed to yet another false binary determining superiority and inferiority.

Over the last decade this conflict has been re-opened by the psychologist Simon Baron-Cohen, who returned to the hormonal binary, celebrating the 'essential difference' between men and women by drawing on 'brain organisation theory'.[21] The essential difference, he argues, results from miniscule differences in brain structure, encoded in the genes and fixed by a 'testosterone surge' during the third trimester of the male's foetal life, which is supposed to result in women's superiority in linguistic and emotional skills, and men's in spatial and systematising abilities.

Feminist neuroscientists Rebecca Jordan-Young[22] and Cordelia Fine[23] have been among those taking on this latest manifestation of masculinist essentialism. Fine, a psychologist, subjects a range of neurosciences to sharp methodological criticism, dismissing their essentialist and sexist conclusions. Jordan-Young's target is brain organisation theory. Grounded in feminist science studies, she critiques the methodology employed by brain organisation theory researchers but more radically rejects the simple nature/culture binary on which essentialism relies. Without rehearsing a long theoretical debate, today's feminist theorists, whether from the life or social sciences, increasingly see these boundaries of nature/culture as under continuous negotiation. Certainly neither biology nor sociology can have the last word.

CAN BIOETHICS BELL THE BIOTECHNOSCIENCE CAT?

Large-scale experimentation on humans began early in the last century, with research – usually public health or military – on the learning disabled, prisoners and others lacking social power. For the Nazi doctors the justifications that they were doing good science and that their subjects were by nature *untermenschen* were all too close kin to those made by public health researchers in Tuskegee, Guatemala and elsewhere. Such research, always partly shaped by the interests of the researchers themselves, was driven by the state's interest in maintaining a healthy and productive civilian population and a fit and effective military; only later did it come to serve the financial interests of a burgeoning biomedical and pharmaceutical industry.

Whether the origins of bioethics as both discipline and institution lie in the Nuremberg Doctor's Trial, as in the US version of its history, or in response to the scandals exposed by the US civil rights movement, they share a fundamental commitment to the ethic that research subjects must be able to act voluntarily, free to give or withhold their consent without coercion or penalty. US bioethicists' concept of informed consent held until the immense changes in the production of biomedical knowledge with the advent of human genomics. As bioethicist Arthur Caplan charged, if an informed-consent form for gene therapy runs to fifteen pages, informed consent is 'broken' and needs fixing.[24] The problem is not only the fix but who is to do the fixing.

It was the 5 per cent of the HGP budget committed to its Ethical Legal and Social Implications (ELSI) that transformed bioethics. The money available was huge, wildly disproportionate to the available pool of experts and expertise. Low-quality research, vacuous conferences, unreadable reports and specialised journals grew like weeds, and out of all of this lushly funded confusion rose the academic enterprise that bioethics has become today.

These new experts have, as US sociologists Erin Conrad and Raymond de Vries observed, 'successfully insinuated [themselves] into the infrastructure of medicine and the life sciences'.[25] *Nature*, however, read this development approvingly, commenting that university biomedical researchers could now 'dial E for Ethics' and consult a bioethicist on campus.[26] Being on the same campus facilitates the necessary dialogue between the scientists and bioethicists, but carries the danger of closeness and loss of independence for the bioethicists. When both scientist and ethicist are paid from the same research budget, which can and does happen, the problem of independence only becomes more complicated.

Today all research involving human subjects in Europe, and all Federally funded research in the US, must go through formal ethical review. Biotechnology companies regularly appoint ethics advisory boards. International agencies, most notably UNESCO, have long had bioethics committees. New subfields – genethics, neuroethics – have spawned, their founders claiming that each raises distinct ethical problems, requiring specialised expertise and dedicated funding. As Conrad and de Vries concluded, 'Bioethics has clearly succeeded in claiming an area of expertise and making its work indispensible to medicine and medical research.' The consequence has been that moral consideration of the issues and the potential consequences of biomedical research can be left to others, 'outsourced' to qualified experts.

The urgent imperatives of the technosciences within globalisation fostered the institutionalised outsourcing of the social and moral responsibility of the scientists and technologists. While scientists have no special ethical responsibility over and above that of any other citizen, they do have expert knowledge of science and science-related risk, and thus have a specific part to play. As outsourced bioethics becomes ever more closely incorporated into the global biotech economy, the chances of it helping restrain biotech's activities diminish. Modern science's claim to reason, rationality and objectivity has long sought to exclude love, responsibility and indeed all moral sentiment. Faced with ethical problems associated with its research, science's institutional response has been to wash its hands: science, it claims, is neutral.

As sociologist Zygmunt Bauman points out, 'the most prominent among the exiled emotions within globalisation are moral sentiments'.[27] In the sense that the telos of globalisation and its crises is to enrich the few at the expense of the many, Bauman is surely right. But where there are adequate resources moral thinking and feeling can still flourish. This was elegantly demonstrated by Rayna Rapp's study of women in New York's antenatal clinic being confronted by a proliferation of diagnostic tests, working out what these meant for the potential baby, their families and themselves.[28] The autonomous, cognitively driven individual of principlist bioethics was nowhere to be seen; instead the women practised a relational ethics in order to find their way. Given the magnitude of the new biomedical technologies they faced, Rapp rightly calls these women 'moral pioneers'. Here we see a bioethics from below in contrast to the bioethics from above of the professionals.

Similarly, the DNA biobank established and controlled by the Little People of America, a self-governing association of people with achondroplasia, offers not only a bioethics from below but also effective regulation. Fed up with genetic researchers constantly requesting blood samples, whose connection to their problems often seemed remote if not non-existent, they decided to take control. Now only those genetic researchers whose projects they trust get approval and secure access to the biobank.

Not all patients' groups enjoy such power, however. Financially independent patient groups are the exception; many similar groups are both fostered and funded by Pharma, at worst deliberately recruited in the companies' lobbying campaigns for increased research funding and regulatory approval for novel and expensive drugs. Behind closed walls, with only professionals present, senior representatives from the industry unblushingly confirm the practice.

GOODBYE TRUTH: HELLO TRUST?

The last quarter of the twentieth century saw a remarkable transformation in the relationship between 'science' and 'society'. Science's claim that it had the final word in the naming of nature, that its claims to truth were stronger than any others, began to weaken, challenged from below by the new social justice movements and their intellectuals, and subverted by the fast expanding field of science studies.

The culture of deference towards biomedical experts as the guardians of truth showed cracks in the 1970s, with the feminist movement's alternative biosocial construction of women's bodies. *Our Bodies, Ourselves* was both handbook and icon. This collectively produced knowledge with its heretical egalitarian epistemology set aside formal expertise, instead beginning with the shared personal experiences of living within a woman's body. Many rejected modern biomedicine altogether; sharing Virginia Woolf's hostility to patriarchal science, they turned to homeopathy and similar alternative therapies.

Where science had served to disenchant the world, the growth of the multidisciplinary project of the social studies of science within academia has served to disenchant science itself. The subversive move of science studies was to treat scientific knowledge as any other knowledge. Laboratories were just new places and scientists just new groups of people to study as they went about their work of building scientific knowledge. Anthropologist Bruno Latour study of a group of biomedical researchers at the Salk Institute in California, and their construction of scientific 'facts', *Laboratory Life*, the book he subsequently wrote together with sociologist Steve Woolgar, epitomised this move.[29]

Truth, and hence trust, in science suffered a series of self-inflicted wounds. In Europe there was a mass rejection of GM food. The assurances from science, government and industry that the GM food was safe didn't wash. The prioritisation of commercial interests over public safety was for most all too evident. The British in particular were hit by a plethora of scandals: outrage at the governmental mismanagement of Mad Cow disease (BSE), the Alder Hey pathologist secretly removing organs, and the child heart surgeon in Bristol carrying out operations with lethal incompetence. The patronising public health response to the anxieties felt by parents following a study published in *The Lancet* claiming that the MMR triple vaccine was associated with autism (subsequently withdrawn) didn't help either. The authorities' failure to communicate saw an increased refusal by parents, particularly mothers, to have their children vaccinated, potentially increasing the risk of reduced herd immunity.

But for some scientists, any questioning of their exclusive cultural authority to speak the truth about nature was intellectually and probably psychologically more than they could cope with. Their response, the opening salvo in what became known as the Science Wars, was launched in the US in the mid 1990s with the publication of *The Higher Superstition: The Academic Left and its Quarrels With Science*.[30] Its contributors attacked the critical discussion of science by the humanities and the social sciences, giving their most hostile attention to feminists, with standpoint theorist Sandra Harding their chief target.

For them, science, and science alone, spoke the truth. Where initiatives of public engagement sought to transform the relationship of science and society, the Science Wars marked the symbolic last stand of the ideologues of scientism against their imagined enemy of science studies. What they read as anti-science was part of a change in culture, in which science and the social studies of science no longer talked past one another, but – optimistically – to one another, with the shared aim of renegotiating the relationship between science and society. The debate has not entirely disappeared, however, and the practice of recruiting social scientists, bioethicists and others as handmaidens to secure the acceptability of already planned research has not entirely retreated.

The scientific community's grappling with the critical question of trust marked this slow transformation. Even before the scandals of the '90s began to rain down in full force, the Royal Society saw the need to take seriously this growing threat to the authority of scientific knowledge (and rather more pertinently to the potential loss of public financial support for research). They saw the problem as one of public ignorance, to be remedied by scientists explaining their science to the world more clearly, with the anticipated happy result that their narrative of truth would restore trust.

This cause was taken up with tremendous enthusiasm by academic scientists who enjoyed talking about their research to interested lay audiences. Meanwhile, the Research Councils took the view that public understanding was no different from public relations, with the result that overnight, to the amusement of some of their science journalist staff, their Public Relations units were renamed Public Understanding units, with no refocusing needed.

Under this umbrella came the uncritical celebrations of science, and rather less well-funded critical responses. Science festivals, first pioneered in Edinburgh in 1991 as a younger sibling to the city's arts festival, spawned like mushrooms, as did Cafés Scientifique in cities across Europe and America. The sedately titled Annual Meeting of the British Association for the Advancement of Science rebranded itself as the British Festival of Science, though it remained largely uncritical, as did the EU with its annual Science Awareness Week.

Unfortunately, the 1990s scandals had demonstrated that Public Understanding was not enough. To get the public's trust back, the establishment needed to recognise that, on issues which directly affected the public, a cultural change in science itself was required. The public could no longer be seen as an empty vessel passively waiting for scientific truth to be poured in. Instead of a single homogeneous group, 'the public' had to be recognised as consisting of plural and diverse publics. To maintain trust in policy advice on science-related issues, governments and scientists increasingly recognised that while the scientific experts were still needed, it was a mistake to exclude lay expertise. The monologue of the expert had to yield to dialogue with the public.

The incoming 1997 Labour government was already conscious of the need to bring non-experts into the science advisory process. As mentioned in the Introduction, they built on the previous governments' innovation of the HFEA, adding the Food Standards Agency (like the HFEA with regulatory powers) and two further standing Commissions with advisory status, for human genetics and for agriculture and biotechnology, each replete with lay representatives. For some of the older hands in science studies, the largely failed 1960s project of democratising science looked to be making a return, but this time equipped with a more sophisticated understanding of the science-and-society relationship and a renewed interest in the new social technologies of public participation.

PUBLIC ENGAGEMENT

The scientific elite of the mid-twentieth century had disapproved of their juniors speaking to the public, as a distraction from their true calling and a loss of valuable time away from the laboratory bench. Now, however, the Research Councils and the Royal Society specifically encouraged public dialogue and began to train young scientists in science communication.

A new kind of scientist appeared, more comfortable in dialogue with the diverse publics, no longer penalised by loss of professional esteem. This new breed treated with respect the contributions of participating citizens, who in turn visibly enjoyed both learning about science and asking penetrating questions. Nor were the participants politically passive; in one Welsh citizen's jury discussing the social implications of the new human genetics, the jurors criticised the absence of black Welsh participants. The organiser's reply that statistically there were insufficient numbers to justify their participation cut little ice. The participants' view, as taken-for-granted multiculturalists, was that Welsh people from different cultures should all have their say was more important than being dictated to by the percentages.

One problem that bedevils public consultation is the choice of which social technology to choose to ensure adequate representation of the increasing diversity among the European population. Should it be a representative group, or is it more fruitful to recruit lay people with relevant expertise derived from their everyday lives? The powerful credo of the disability movement – 'nothing about me without me' – suggests that to leave such lay experts out is a mistake.

A major European consultation exercise carried out in 2004 was titled *The Meeting of Minds: The Citizens' Deliberation on Brain Research* – as if 'mind' was unquestionably coterminous with 'brain'. The consultation, EU-funded but organised by the King Baudouin Foundation, sought to learn what 'the people of Europe' believed should be the priorities for brain research and to make recommendations to the EU commission and parliament as to the issues that needed vigilance. The project was among the most ambitious and best-funded of its kind; the 'people of Europe' were to be represented by a panel of 126 drawn from nine countries. Given its claims to speak for European diversity, however, the fact that Steven as a participating neuroscientist could see no one of colour on the entire panel was, to say the least, disturbing. Given the very different experience of black and ethnic minorities using mental health services, and their known hostility to taking part in research, their absence must have had implications for the final recommendations.

The citizens met in both national and international groups, to interrogate the neuroexperts and, at a final two-day plenary in Brussels, to formulate recommendations. But on what basis could the recommendations be made? Among the experts and organisations offering advice and opinions were individual neuroscientists, the European Brain Council – a lobby group for increased funding for brain research – and representatives from Pharma. Conspicuous by their absence were critics of biological psychiatry, despite the evidence that whatever it was doing it wasn't enough to stop either the revolving doors of the psychiatric wards or the growing numbers of the mentally ill. There was little background information provided – such as how much money was going into European brain research or the balance between research and care – making informed judgement by the citizens that much more difficult.

Energetic facilitators at the final plenary ensured a list of consensus recommendations including more resources for the brain researchers, a public education project to contest stigmatisation, a greater input by lay people into governance, and the strengthening of ethical oversight. The practical outcomes for a Europe-wide de-stigmatisation campaign and the increased upstream lay input into the European research system are still unclear, but the well-organised European Brain Council has continued to challenge what they see as the underfunding of brain research relative to heart and cancer. Lacking a similar organisational base 'society' has a poor chance of increasing its voice – except when politicians are compelled to listen by political protest or moral crisis, as was the case with Europe's rage against GM.

In contrast to public engagement conceived as a more democratic way of shaping science policy one finds the fierce debate among Iceland's clinicians, geneticists, ethicists and the general public, played out in parliament and in the press, over deCode's nationwide biobank. This was not engagement but self-funded opposition, relying on mobilisation on ethical, political and scientific grounds. The political judgement of two anthropologists, the American Paul Rabinow and the Icelandic Gisli Palsson,[31] that the pros and antis were more or less equal in strength seems a little curious when a shoestring David (albeit with international biomedical allies) confronted a well-funded Goliath supported by a neoliberal government.

Nonetheless the opt-out campaign, focusing on the issue of

informed consent and the commodification of the Icelanders' bio-information, succeeded – as confirmed by the ruling of the appeal court that the compulsory inclusion of records concerning the dead and minors was unconstitutional. The dramatic process served almost incidentally as a mass education project for the Icelanders in the ethical social and legal implications of DNA biobanking. Unfortunately a lesson not well learnt elsewhere.

The intensity of the national debate in Iceland, and the lesser but still vigorous Swedish debate, contrast all too vividly with the sedate discussions of UK Biobank. Parliamentary scrutiny was restricted to a Commons Select Committee so that although it raised the question of the opportunity costs of putting so much of the research funding into the biobank, the hinted-at full parliamentary debate never took place. Worse still, the referees' reports asked for by the Select Committee were extracted only after a freedom of information request by *GeneWatch*. Even the sharp doubts expressed in the UK's leading medical journals and the criticisms of independent geneticists never generated a public debate anywhere near that which took place in Iceland.

UK Biobank did indeed proceed slowly with several consultation exercises to build trust. One obvious measure of success in winning public trust is the participation rate, that is, the percentage of people invited to take part who agree to do so. It was known that participation rates in genetic population studies had been going down in recent years, tending to vary between 60 and 40 per cent, as some of UK Biobank's referees pointed out. The original protocol envisaged a modest 25 per cent, but the response rate in the pilot studies was a miserable 10 per cent. While the biobank's proponents took the view that this did not affect a DNA population project, others raised concerns about its representativeness given the UK's highly diverse population and the well-documented refusal to participate of people marginalised by their socio-economic and/or minority status.

Was UK Biobank misled by the reassurance it received from the social research it had commissioned? Or was the low participation rate an indication of a lack of popular trust? The resistance of the biobank's planners to ruling out the possibility of insurance and tobacco companies having access to the samples and data made some consultees distinctly uncomfortable. Some of UK Biobank's prominent backers appealed to the altruism of participating in a biomedical project that aimed both to provide better explanations of disease and to open the way to personalised medicine. The 90 per cent refusal rate suggests, however, that what UK Biobank spoke of as an opportunity for altruism – the population's gift of their medical records and genes – was not understood as such by the great majority of the invitees.

ALTRUISM FOR SOME?

The name of the distinguished English social policy analyst Richard Titmuss has been repeatedly invoked in the cause of persuading citizens to donate both their personal health information and their tissue to biobank research. It was Titmuss's 1970 study of the UK blood donation service that led to his celebrated discussion of altruism as the gift to the stranger.[32] Despite Titmuss locating his thesis in the context of anthropological theories of the gift, he misunderstood the literature. As Cambridge anthropologist Edmund Leach generously observed: 'Titmuss' anthropology may be a bit shaky but his moral case is founded on rock.'[33]

The altruism of giving blood to help an unknown patient (rather than a biomedical research project) was for Titmuss integral to the solidarity of the welfare state. His concept of altruism was rooted in the unconditional security offered by the National Health Service – for Titmuss always the jewel in the welfare state crown. He shared and advocated the collectivist consensus that the provision of health care was to be determined by need not income, and pointed out that when blood is commodified, as in the US where it is bought and sold, its quality can be no longer guaranteed. For blood to be safe, its status as a gift, falling outside of commodity relations, must be celebrated and defended.

Titmuss's notion of altruism was not some free-floating concept unrooted in peoples' lives, but was embedded in mutual collectivism. Listening to his reflections on waiting in the cancer clinic

– to be treated for the cancer which was to kill him – he speaks of the patient who was to be seen next by the clinician: this was not decided by wealth or social status, the university professor before the bus conductor, but by the vagaries of the London traffic which determined who arrived first.

With the possibility of human embryonic stem cells as the key to regenerative medicine, researchers turned initially to women who, following IVF had 'surplus' embryos as potential providers. There was already a precedent: women did share eggs with other women, an altruistic act sometimes encouraged (incentivised?) by the offer of quicker treatment. This regulated world of the past is now challenged by the footloose ethics of global capitalism. Those with the money can always turn to reproductive tourism, to countries with laxer regulations and lighter costs. Cyprus has become a focal point for the global human egg trade, in which the fees paid for poor women's eggs are less than for those from rich women – Ukrainian and Russian eggs being among the cheapest. Israeli-run clinics collect eggs to maintain the IVF rate in Israel, already the highest the world, richly subsidised as part of the demographic race against the Palestinians. If we take the Titmuss thesis seriously, just how reliable can quality control be when eggs are bought and sold?

The shortage of eggs means that, even when a couple have given their informed consent, researchers are likely to get rejects deemed not good enough to implant. Hence, as we saw with the Hwang story, scientists will become intensely interested in securing high-quality eggs produced for research not reproduction. Social scientists originally found that most women undergoing IVF felt that providing eggs should be a gift rather than a financial transaction and were more comfortable with the idea of donating them for reproduction than for experiment. More recent work turns this about: more women are now reported as in favour of financial compensation and as open to the idea of providing eggs for hESC studies. Is this shift part of the wider cultural shift towards the values of neoliberalism, suggesting a deepening alienation from ourselves and our increasingly commodified bodies?

The Nuffield Council's report discussing 'donation' interrupts the Titmuss-inspired call for altruism made by the bioethicists on very different grounds than we have made here.[34] They argue that donating eggs to another woman is very different from providing eggs for hESC research. They regard the first as primarily an altruistic act and the second as more instrumental, where payment would be appropriate. Although women, including feminists, were strongly represented on the committee, feminist arguments against women's bodies being mined to secure research resources were not represented. Instead, the mortality risk (0.05 per cent) to the egg providers was compared with the risk to healthy volunteers in the first phase of clinical trials. This only produces a scenario in which neither secular nor religious oppositionists are likely to be swayed. Although there has to be some relief that pro-hESC meetings may no longer parade women who have donated eggs for research as advocates for altruism, raising that nineteenth-century spectre of the Angel in the House, the endlessly self-sacrificing woman, her altruism incessantly praised while the injustices done to her are discounted, the skewed morality of altruism for women but not for men reappears with hESC. This is well illustrated by a comment on a new hESC proposal by Professor of Reproductive Medicine and prominent hESC researcher Alison Murdoch:

> Many women have expressed an interest in helping because they have family or friends who have been affected … or because they are interested in helping the research. Until now, we have not been able to recruit non-patient donors but that has now changed. After receiving counselling, these women are able to donate eggs altruistically – and these donations will be vital in providing a source of eggs for the researchers to be able to take forward their work towards eliminating these currently incurable diseases.[35]

THE OWL OF MINERVA

Hegel observed that wisdom, like the owl of Minerva, only flies at dusk, and so it is with the latest turn in the wisdom of bioethics, invoking 'solidarity' when the political, ethical and practical preconditions for such solidarity have already fled. Solidarity, a recent report for the Nuffield Council claims, has two key dimensions, as moral value and as practice.[36]

The moral value of solidarity supports public participation in large-scale biomedical research, such as providing biosamples and health and life-style data. But how is the moral value of solidarity to be achieved in practice? The collectivist practices of the welfare state have been steadily eroded over the last three decades, from the collapse of the once archetypical Swedish welfare state to the most recent British Attitude Survey reporting public resistance to the very idea of solidarity. The narrowing of the gap between rich and poor which took place while the welfare state was strong has been replaced today by extreme wealth for some and an intensifying precarity for most. Scroungermania, fostered by the concord between the tabloid press and the political classes, has done its work. In this context, exhortations to solidarity by bioethicists are no substitute for new political visions.

This increased reluctance to participate altruistically in large-scale biomedical research, to give without material reward, is expressive of a consumer society in which commodity relationships have replaced collectivism. Bio-solidarity is not dead in so far as blood is still donated, but this is commonly understood by donors as a gift for the purpose of care, rather than research. Talking up consumer choice, an increasingly statist authoritarianism has rowed back from the upstream public engagement in science policy proposed by the House of Lords 2000 report and the European experiments in consultation, with their commitment to transforming the relationship between science and society. Thus in spring 2012 the HFEA announced a public consultation concerning the social and ethical issues raised by a research project submitted by Newcastle University's Mitochondrial Research Unit requiring the use of hESC and therefore needing formal approval. The nucleus of the eggs of the at-risk woman wanting a child free from mitochondrial disease would be inserted into an hESC donated by another woman. The media immediately spotted the core ethical issue, and greeted this as a 'three parent' conception. However, the public engagement only took place after the Wellcome Trust had given £4.6 million to the Research Institute – not so dissimilar to Wellcome's track record with UK Biobank. Scarcely upstream public engagement.

Venture capitalists, leading academic scientists and Big Pharma, uninterested in fluffy talk about solidarity with its doubtful capacity to increase participation rates, argue instead for automatic access to electronic medical records. To the Panopticon already provided to the state by CCTV cameras, and to commerce by Google and supermarket credit cards, comes the recording of the most intimate details of our physical and psychic lives – DNA, health records, neuro-images – as a resource for data mining by the public and commercial sectors alike.

Since the banking meltdown of 2007–8, the neoliberal leaders of Europe and the US are agreed that the welfare of the majority, above all the most vulnerable, must be replaced by welfare payments to bankers. That welfare payments bring psychological security to their recipients is nowhere better demonstrated than by the lack of suicides and the descent into poverty among the present generation of incompetent greedy bankers compared to those of the crash of 1929. In their relentless demand for yet further handouts they show every sign of the welfare dependency they so deplore in the poor.

The exception to this bleak narrative is tiny Iceland, where the story of DNA biobanking began. The combination of political, ethical and scientific opposition consolidated by the ruling of the Supreme Court destroyed the commercial project of a nationwide electronic medical record system linked to the nation's genome. The opposition had taken on and defeated both deCode and its ally the neoliberal government led by David Oddsson, who had publicly declared himself willing to sweep away any ethical constraints that might impede the advance of the new technology. Meanwhile the unrestrained speculation of the Icelandic banks continued, with Oddsson now heading Iceland's central bank.

The banking crash, followed by the usual proposal that the people should bear the costs of the politicians' and bankers' actions, was met by a democratic upsurge. A new constitution was drawn up by randomly selected citizens, and a truth committee appointed which laid the blame at the door of four politicians, including Oddsson and Geir Haarde, Prime Minister at the time of the crash. Iceland's parliament decided to prosecute Haarde for gross negligence, and in April 2012 he was found guilty on one of the charges against him.

There is a lesson here.

Acknowledgements

Our acknowledgements must begin with an apology to those many colleagues and friends on whose work we have drawn but whom we have not cited. We are however profoundly conscious of our debts to you all. At our Verso editor Sebastian Budgen's suggestion, we agreed to keep the references to a minimum to enable the book to reach a wide audience – the same objective we had when we wrote our first joint book, *Science and Society*, back in 1969. Then it was relatively easy; there wasn't a large literature, and most of the people writing we knew personally. Today the literature is vast and the task is correspondingly harder. Our sources range from the pamphlets of the past (now nearly obsolete technologies) to the websites, podcasts, and blogs of today, and to the host of reports, books, journals, and conferences of academic science and technology studies. To do justice to this academic and movement literature would have made the book knee-deep in references.

Here, therefore, we only directly acknowledge those who have read and commented on specific chapters – Patrick Bateson, Jack Price, Jerry Ravetz, Simon Rose, and Helen Wallace – and a particular thanks to the meticulous copy-editing of Tim Clark. Any inaccuracies and infelicities that remain are of course our responsibility. Earlier versions of some chapters have appeared previously. Chapter 2 draws on two joint papers, 'The Changing Face of Human Nature', *Daedalus*, Summer 2009, pp. 7–20, and 'Darwin and After', *New Left Review* 63, May–June 2010, pp. 91–114. The history of deCode in Chapter 5 is updated from Hilary's Wellcome Trust–funded ethnographic study (*The Commodification of Bioinformation:*

The Icelandic Health Sector Database. Her Umangenomics study in Chapter 6 was funded by Gresham College. The discussion of the neurotechnologies in Chapter 8 builds on Steven's work beginning in 2006 for the Navigator Network of the New Zealand Royal Society at the invitation of Barbara Nicholas, and subsequently the Science Policy unit of the UK Royal Society, including its BrainWaves project, which culminated in 2012. The themes in the book began to be developed during our three-year joint tenure as the Gresham Physick Chair in Genetics in Society and consolidated in our Ralph Miliband lecture 'Biotechnology, Globalisation and Democracy', given at the London School of Economics. Hilary would also like to acknowledge the hospitality offered by Director Nikolas Rose and Deputy Sarah Franklin as a guest of the Bios Centre while a Visiting Professor of Sociology at the London School of Economics, 2008–12.

Notes

INTRODUCTION: PROMETHEUS UNBOUND?

1 Rose, H. and Rose, S., *Science and Society*, Allen Lane, 1969.

2 Hessen, B., in Bukharin, N., *Science at the Crossroads* (1931), Cass reprint, 1971.

3 Bernal, J.D., *The Social Functions of Science*, Routledge, 1939.

4 Lewontin, R. and Levins, R., 'The Problem of Lysenkoism', in H. Rose and S. Rose (eds.), *The Radicalisation of Science*, Macmillan, 1976, pp. 32–64.

5 Bernal, J.D., *Marx and Science*, Lawrence and Wishart, 1952.

6 Rose, S. (ed.), *Chemical and Biological Warfare*, Harrap, 1968.

7 Ciccotti, G., Cini, M., and de Maria, M., 'The Production of Science in Advanced Capitalist Society', in Rose, H. and Rose, S. (eds.), *The Political Economy of Science*, Macmillan, 1976, pp. 32–58.

8 Lévy-Leblond, J-M, 'Ideology of/in Contemporary Physics', in Rose and Rose, *The Radicalisation of Science*, pp. 136–75.

9 Couture-Cherki, M., 'Women in Physics', in ibid., pp. 65–75; Stéhelin, L., 'Sciences, Women and Ideology', in ibid., pp. 76–89.

10 Ann Arbor Science for the People Editorial Collective, *Biology as a Social Weapon*, Burgess, 1977.

11 Tobach, E. and Rosoff, B., *Genes and Gender I*, Gordian, 1978 (and later volumes); Arditti, R. (ed.), *Science and Liberation*, South End Press, 1980; Hubbard, R. and Henifin, M.S. (eds.), *Women Look at Biology Looking at Women*, Schenkman, 1979; Bleier, R. (ed.), *Feminist Approaches to Science*, Pergamon, 1986.

12 Shapiro, J., Machattie, L., Eron, L., Ihler, G., Ippen, K., and Beckwith, J., 'Isolation of Pure *lac* Operon DNA', *Nature* 224, 768–74, 1969.

13 Rose, H. and Rose, S., 'Chemical Spraying as Reported by Refugees from South Vietnam', *Science* 117, 710–12, 1972.

14 Brighton Women and Science Group, *Alice Through the Microscope*, Virago, 1980.

15 See Rose, H., 'The Social Determinants of Reproductive Science and Technology', in Knorr, K. and Strasser, H. (eds.), *The Social Determinants of Science, Yearbook of the Sociology of Science*, Reidel, 1975.

16 Anderson, P., 'Components of the National Culture', *New Left Review* 50, 1–37, 1968.

17 Rose, H. and Rose, S., 'Radical Science and its Enemies', *Socialist Register*, 317–35, 1979.

18 Carson, R., *Silent Spring*, Houghton Mifflin, 1962.

19 Beck, U., *Risk Society: Towards a New Modernity*, Sage, 1992.
20 Hartsock, N., *Money, Sex and Power*, Boston University Press, 1984; Harding, S., *The Science Question in Feminism*, Cornell University Press, 1986; Rose, H., 'Hand, Brain and Heart: A Feminist Epistemology for the Natural Sciences', *Signs: Journal of Women in Culture and Society* 9 (3), 73–98, 1983.
21 Shapin, S, *The Scientific Life: A Moral History of a Late Modern Vocation* Chicago UP, 2008.
22 Irwin, A. and Wynne, B., *Misunderstanding Science: The Public Reconstruction of Science and Technology*, Cambridge University Press, 1996.
23 Wilsden, J. (ed.), *See-through Science*, Demos, 2004.
24 Graham, S., *Cities Under Siege: The New Military Urbanism*, Verso, 2010.
25 Ledoux, J., *The Synaptic Self: How Our Brains Become Who We Are*, Viking, 2002; Damasio, A., *Descartes' Error: Emotion, Reason and the Human Brain*, GP Putnam and Sons, 1994.
26 Zeki, S., *Splendours and Miseries of the Brain: Love, Creativity and the Quest for Human Happiness*, Wiley-Blackwell, 2009; Crick, F.H.C., *The Astonishing Hypothesis: The Scientific Search for the Soul*, Simon and Schuster, 1994.
27 Changeaux, J-P., *Neuronal Man: The Biology of Mind*, Princeton University Press, 1997.
28 Baron-Cohen, S., *The Essential Difference: The Truth About the Male and Female Brain*, Basic Books, 2003.
29 Fine, C., *Delusions of Gender: The Real Science Behind Sex Differences*, Icon, 2010; Jordan-Young, R., *Brainstorm: The Flaws in the Science of Sex Differences*, Harvard University Press, 2010.

CHAPTER 1: FROM LITTLE GENETICS TO BIG GENOMICS

1 Gilbert, W.A., 'Vision of the Grail', in Kevles, D.J. and Hood, L. (eds.), *The Code of Codes: Scientific and Social Issues in the Human Genome Project*, Harvard, 1990, p. 96.
2 Watson, J.D. and Crick, F.H.C., 'Molecular Structure of Nucleic Acids: A Structure for Deoxyribose Nucleic Acid', *Nature* 171, 737–8, 1953.
3 Wilkins, M.H.F., *Maurice Wilkins: The Third Man of the Double Helix*, Oxford, 2005.
4 Sayre, A., *Rosalind Franklin and DNA*, Norton, 1975.
5 Maddox, B., *Rosalind Franklin: The Dark Lady of DNA*, HarperCollins, 2002.
6 De Solla Price, D.J., *Little Science, Big Science*, Columbia, 1963.
7 Brown, A., *In the Beginning Was the Worm: Finding the Secrets of Life in a Tiny Hermaphrodite*, Simon and Schuster, 2003.
8 Koshland, D., 'Sequences and Consequences of the Human Genome', *Science* 246, 189, 1989.
9 Cook-Deegan, R., *The Gene Wars: Science, Politics, and the Human Genome*, Norton, 1995.
10 Collins, F., *The Language of God: A Scientist Presents Evidence for Belief*, Free Press, 2006.
11 International Human Genome Sequencing Consortium, 'Initial Sequencing and Analysis of the Human Genome', *Nature* 409, 860–921, 2001.

12 Venter, J.C., *A Life Decoded: My Genome, My Life*, Penguin, 2008.
13 Collins, F.S., *The Language of Life: DNA and the Revolution in Personalised Medicine*, Harper, 2009.
14 Sulston, J. and Ferry, F., *The Common Thread*, Bantam 2002.
15 Cavalli-Sforza, L.L., 'The Human Genome Diversity Project: Past, Present and Future', *Nature Reviews Genetics* 6, 333–40, 2005.
16 Marks, J., *What it Means to be 98% Chimpanzee: Apes, People and Their Genes*, University of California Press, 2002.
17 Apaza, B.M., quoted in *Science*, 332, 773, 2011.
18 Reardon, J., *Race to the Finish: Identity and Governance in an Age of Genomics*, Princeton, 2005.
19 International HapMap Consortium, 'Integrating Ethics and Science in the International HapMap Project', *Nature Reviews Genetics* 5, 467–75, 2004.
20 Battelle Institute, *Economic Impact of the Human Genome Project*, Battelle Technology Partnership Project, May 2011.
21 Waxman, J., personal communication.
22 Quoted in Marshall, E., 'Waiting for the Revolution', *Science* 331, 526–9, 2011.
23 Lander, E.S., 'Initial Impact of the Sequencing of the Human Genome', *Nature* 470, 187–203, 2010.
24 Dawkins, R., *The Blind Watchmaker*, Norton, 1986.

CHAPTER 2: EVOLUTIONARY THEORY IN THE POST-GENOMIC AGE

1 Carroll, S., *Endless Forms Most Beautiful*, Weidenfeld, 2006.
2 From Moleschott's 1852 text *Das Kreislauf des Lebens*, quoted in Donald Fleming's Introduction to Jacques Loeb's *The Mechanistic Conception of Life* (1912), Bellknap Press, 1964.
3 Darwin, F. (ed.), *The Autobiography of Charles Darwin*, Dover, 1958, pp. 43.
4 Wallace, A.R., *Social Environment and Moral Progress*, Cassell, 1913, pp. 147–8, cited by Benton, T., 'Social Causes and Natural Relations', in Rose, H. and Rose, S. (eds.), *Alas Poor Darwin*, Cape, 2000, pp. 249–71.
5 Marx, K. and Engels, F., *Collected Works*, Vol. 41, Moscow, 1985, p. 280.
6 Darwin, quoted in Hubbard, R., 'Have Only Men Evolved?', in Hubbard, R. and Henifin, M.S. (eds.), *Women Look at Biology Looking at Women*, Schenkman, 1979, pp. 19–20.
7 Blackwell, A.B., *The Sexes Through Nature*, G.P. Putnam and Son, 1875.
8 Bateson, P.P.G. and Gluckman, P., *Plasticity, Robustness, Development and Evolution*, Cambridge University Press, 2011.
9 Needham, J., *Time the Refreshing River*, Macmillan, 1943.
10 Haraway, D., *Crystals, Fabrics and Fields*, Yale, 1976.
11 Dobzhansky, T., 'Nothing in Biology Makes Sense Except in the Light of Evolution', *American Biology Teacher* 35, 125–9, 1973.
12 Dawkins, R., *The Selfish Gene*, Oxford University Press, 1976.
13 Wilson, E.O., *Sociobiology: The New Synthesis*, Harvard University Press, 1975.

14 Ann Arbor Science for the People Editorial Collective, *Biology as a Social Weapon*, Burgess, 1977.

15 Lee, R. and DeVore, I., *Man the Hunter*, Aldine, 1968.

16 Zihlman, A.L., 'Women as Shapers of the Human Adaptation', in Dahlberg, F. (ed.), *Woman the Gatherer*, Yale University Press, 1981, pp. 75–120.

17 Haraway, D., *Primate Visions: Gender, Race and Nature in the World of Modern Science*, Routledge, 1989.

18 Jolly, A. and Jolly, M., 'A View From the Other End of the Telescope', *New Scientist*, 21 April 1990, p. 58.

19 Keller, E.F., *The Century of the Gene*, Harvard University Press, 2002.

20 Oyama, S., *The Ontogeny of Information: Developmental Systems and Evolution*, Cambridge University Press, 1985.

21 Maturana H.R. and Varela F.J., *The Tree of Knowledge: The Biological Roots of Human Understanding*, Shambhala, 1998.

22 McLaren, A., Personal communication.

23 Rose, S., Lewontin, R.C., and Kamin, L., *Not in Our Genes*, Allen Lane, 1984.

24 Jablonka, E. and Lamb, M., *Evolution in Four Dimensions*, MIT Press, 2005.

25 Miller, G., 'The Seductive Allure of Behavioural Epigenetics', *Science* 329, 24–7, 2010.

26 Bell, A.M. and Robinson, G.E., 'Behavior and the Dynamic Genome', *Science* 332, 1161–2, 2011.

27 Gould, S.J. and Lewontin, R.C., 'The Spandrels of San Marco and the Panglossian Paradigm: A Critique of the Adaptationist Programme', *Proc. Roy. Soc. B* 205, 581–98, 1979.

28 Gould, S.J., *Wonderful Life: The Burgess Shale and the Nature of History*, Penguin, 1989.

29 Conway Morris, S., *Life's Solution: Inevitable Humans in a Lonely Universe*, Cambridge University Press, 2003.

30 Astrobiology, *Philosophical Transactions of the Royal Society, A*, 369, 2011.

31 Dunbar, R., *How Many Friends Does One Person Need?*, Faber, 2010.

32 Dennett, D., *Darwin's Dangerous Idea*, Allen Lane, 1995.

33 Richerson, P. and Boyd, R., *Not by Genes Alone: How Culture Transformed Human Evolution*, Chicago University Press, 2005.

34 Sober, E. and Sloan Wilson, D., *Unto Others: The Evolution and Psychology of Unselfish Behavior*, Harvard University Press, 1998.

35 Nowak, M. with Highfield, R., *SuperCooperators: Altruism, Evolution and Why we Need Each Other to Succeed*, Free Press, 2011.

36 Wilson, E.O., *The Social Conquest of Earth*, Norton, 2012; Dawkins, R., 'The Descent of Edward Wilson', *Prospect*, May 2012.

37 Shippey, T., 'Widowers on the Prowl', *London Review of Books* 33:6, 33–5, 17 March 2011.

38 Hauser, M.D., *Moral Minds*, HarperCollins, 2006.

39 Editorial, *Nature*, 466, 1023, 2010.

40 Wilson, E.O., *On Human Nature*, Harvard University Press, 1979.

41 Pinker, S., *How the Mind Works*, Allen Lane, 1998.

42 Cited by Judith Masters, *The Women's Review of Books* 7:4, 19, 1990.

CHAPTER 3: ANIMALS FIRST: ETHICS ENTERS THE LABORATORY

1 Cobbe, F.P., *Life of Frances Power Cobbe: As Told by Herself*, posthumous edition, London, 1904.

2 Bernard, C., *Introduction a l'étude de la Médicine Expérimentale*, 1865.

3 Moreno, J.D., *Undue Risk: Secret State Experiments on Humans*, Routledge, 2001.

4 Muller Hill, B., *Murderous Science: Elimination by Scientific Selection of Jews, Gypsies, and Others – Germany, 1933–45*, Oxford, 1988.

5 Excerpts From Trial of War Criminals Before the Nuremberg Military Tribunals Under Control Council Law. No/10. October 1946–April 1949. Washington DC: US, GPO, 1949–53.

6 Cited by Evelyne Shuster in 'Fifty Years Later: The Significance of the Nuremberg Code', *New England Journal of Medicine* 337, 1436–40, 1997.

7 Ibid.

8 Holland, J.F., 'The Krebiozen Story', Quackwatch; available at www.quackwatch.com.

9 Moreno, *Undue Risk*.

10 104th Congress, 2nd session, resolution 69. 'Expressing the sense of the Congress that the German Government should investigate and prosecute Dr Hans Joachim Sewering for his war crimes of euthanasia committed during World War II', 2 August 1996.

11 Watson, P., *War on the Mind*, Hutchinson, 1978.

12 *Final Report of the Advisory Committee on Human Radiation Experiments*, Oxford University Press, 1996.

13 Davison, N., *'Non-lethal' Weapons*, Palgrave Macmillan, 2009.

14 Evans, R. and Bowcott, O., 'Veterans Close to MOD Deal', *Guardian*, 18 January 2008.

15 Webb, H.E., Wetherley-Mein, G., Smith, C.E., and McMahon, D., 'Leukaemia and Neoplastic Processes Treated with Langat and Kyasanur Forest Disease Viruses: A Clinical and Laboratory Study of 28 patients', *British Medical Journal* 1, 258–66, 1966.

16 Jones, J., *Bad Blood: The Tuskegee Syphilis Experiment*, Free Press, 1993, p. 179.

17 Skloot, R., *The Immortal Life of Henrietta Lacks*, Macmillan, 2010, p. 135.

18 Barber, B., Lally, J., Makaruschka, J.L., and Sullivan, D., *Research on Human Subjects: Problems of Social Control in Medical Experimentation*, Sage, 1973.

19 Mark, V.H. and Ervin, F.R., *Violence and the Brain*, Harper and Row, 1970.

20 Duster, T. *Backdoor to Eugenics*, Routledge, 1990.

21 Interview with Susan Reverby, *Nature* 467, 645, 2010.

22 Fox, R.C. and Swazey, J.P., *Observing Bioethics*, Oxford, 2008.

23 *The Belmont Report: Ethical Guidelines for the Protection of Human Subjects of Research*, DHEW Publication (OS) 78–0014, Washington, 1978.

24 Beauchamp, T.L. and Childress, J.F., *Principles of Biomedical Ethics*, Oxford, 1994.

25 Annas, G.J., *Standard of Care: The Law of American Bioethics*, Oxford, 1993.

26 Pilcher, H., 'Dial E for Ethics', *Nature* 440, 1104–5, 2006.

27 DeVries, R. and Subedi, J. (eds.), *Bioethics and Society: Constructing the Ethical Enterprise*, Prentice Hall, 1998.

28 Rosenberg, C.E., 'Meanings, Policies, and Medicine: On the Bioethical Enterprise and History', *Daedalus* 128, 27–46, 1999.

29 Tong, R., *Feminist Approaches to Bioethics: Theoretical Reflections and Practical Applications*, Westview, 1997.

30 Donchin, A., 'Autonomy and Interdependence', in Mackenzie, C. and Stoljar, N. (eds.), *Relational Autonomy: Feminist Perspectives on Autonomy, Agency and the Social Self*, Oxford University Press, 2000, pp. 236–58.

31 Ruddick, S., *Maternal Thinking: Towards a Politics of Peace*, Beacon, 1989.

32 Gilligan, C., *In a Different Voice*, Harvard, 1982.

33 Knoppers, B.M. and Chadwick, R., 'Human Genetic Research: Emerging Trends in Ethics, *Nature Reviews Genetics* 6, 75–9, 2005.

34 Dickenson, D., *Property in the Body: Feminist perspectives*, Cambridge University Press, 2007

35 Kaiser, J., 'Gene Therapists Celebrate a Decade of Progress', *Science* 334, 29–30, 2011.

36 Abadie, R., *The Professional Guinea Pig: Big Pharma and the Risky World of Human Subjects*, Duke University, 2010.

37 BBC news channel, 'Six Taken Ill After Drug Trial', 15 March 2006.

38 Petryna, A., *When Experiments Travel: Clinical Trials and the Global Search for Human Subjects*, Princeton University, 2009.

39 Annas, G.J., 'Globalised Clinical Trials and Informed Consent', *New England Journal of Medicine* 360, 2050–3, 2009.

40 Fuller, W. (ed.), *The Social Impact of Modern Biology*, Routledge, 1970.

41 Warnock Report: *Report of the Committee of Inquiry Into Human Fertilisation and Embryology*, Cmd 9314, HMSO, 1984.

42 Wilmut, I., Campbell, K., and Tudge, C., *The Second Creation: The Age of Biological Control by the Scientists Who Cloned Dolly*, Headline, 2000.

CHAPTER 4: FROM STATE TO CONSUMER EUGENICS

1 Keller, E.F., *The Century of the Gene*, Harvard University Press, 2002.

2 Holmes, quoted in Kevles D.J., *In the Name of Eugenics*, Penguin, 1985.

3 Gordon, L., *Woman's Body, Woman's Right: A Social History of Birth Control in America*, Viking, 1976.

4 Pichot, A., *The Pure Society From Darwin to Hitler*, Verso, 2009.

5 Baur, E., Fischer, E., and Lenz, F., *Human Heredity*, Allen and Unwin, 1931.

6 Broberg, G. and Rolls-Hansen, N. (eds.), *Eugenics and the Welfare State: Sterilisation Policy in Denmark, Sweden, Norway and Finland*, p. 138, Michigan State University Press, 1996.

7 Muller, H., *Out of the Night: A Biologist's View of the Future*, Vanguard, 1935.

8 Broberg and Rolls-Hansen, *Eugenics and the Welfare State.*

9 Angastiniotis, M., Kyriakidou, S., and Hadjiminas, M., 'How Thalassaemia Was Controlled in Cyprus', *World Health Forum* 7, 291–7, 1986.

10 Pettit, B., *Invisible Men: Mass Incarceration and the Myth of Black Progress*, Russell Sage, 2012.

11 Titmuss, R.M. and Abel-Smith, B., *Social Policies and Population Growth in Mauritius*, Cass, 1968.

12 Cited in Kay, L., *Who Wrote the Book of Life: A History of the Genetic Code*, Stanford University Press, 1993, p. 276.
13 Humphrey, J., in Watson Fuller (ed.), *The Social Impact of Modern Biology*, Routledge Kegan Paul, 1971, p. 106.
14 Jones, S., *In the Blood: God, Genes and Destiny*, Houghton Mifflin, 1997.
15 Katz Rothman, B., *The Tentative Pregnancy*, Norton, 1993.
16 Rapp, R., *Testing Women, Testing the Fetus: The Social Impact of Amniocentesis in America*, Routledge, 2000.
17 Kitcher, P., *The Lives to Come*, Penguin, 1997.
18 Shakespeare, T., *Disability Rights and Wrongs*, Routledge, 2006.
19 Duster, T., *Backdoor to Eugenics*, Routledge, 1990, p. 92.
20 Ehrlich, P.R., *The Population Bomb*, Ballantine, 1968.
21 Ehrlich, P.R., Ehrlich, A.H., and Holdren, J.P., *Ecoscience: Population, Resources, Environment*, Freeman, 1977.
22 Fausto-Sterling, A., 'The Five Sexes: Why Male and Female Are Not Enough', *The Sciences*, March–April 1993, pp. 20–4.
23 Jordan-Young, R., *Brainstorm: The Flaws in the Science of Sex Differences*, Harvard University Press, 2010.
24 Kevles, *In the Name of Eugenics*, pp. 267–8.
25 Silver, L.M., *Remaking Eden: Cloning and Beyond in a Brave New World*, Weidenfeld and Nicolson, 1998.
26 Thorndike, E.L., *Educational Psychology*, Columbia University Teachers College, 1903, p. 140.
27 Harris, J., *Clones, Genes and Immortality*, Oxford, 1998.
28 Sandel, M.J., *The Case Against Perfection: Ethics in the Age of Genetic Engineering*, Bellknap, Harvard University Press, 2007.
29 Baylis, F. and Robert, J.S., 'The Inevitability of Genetic Enhancement Technologies', *Bioethics* 18, 1–26, 2004.
30 Buchanan, A., *Beyond Humanity*, Oxford University Press, 2011.
31 Kurzweil, R., *The Singularity is Near*, Penguin, 2005.

CHAPTER 5: THE NORTH ATLANTIC BUBBLE

1 Rose, H., *The Commodification of Bioinformation: The Icelandic Health Sector Database*, Wellcome Trust, www.wellcome.ac.uk/stellant/groups/, Public Interest, 2001.
2 Palsson, G., 'The Rise and Fall of a Biobank: The Case of Iceland', in Gottweis, H. and Petersen, A. (eds.), *Biobanks: Governance in Comparative Perspective*, Routledge, 2008, pp. 41–55.
3 Arnason E., 'Personal Identifiability in the Icelandic Health Sector Database', *Journal of Information, Law and Technology* 2, 2002.
4 Indridason, A., *Jar City*, Harvill Press, 2004.
5 Gottweis, H., 'Biobanks in Action: New Strategies in the Governance of Life', in Gottweis and Petersen (eds.), *Biobanks*, pp. 22–38.
6 Leigh, D., *Guardian*, 3 November 2006, p. 15.

7 Kaiser, J., 'deCODE Genetics Rises From the Ashes', *ScienceInsider*, AAAS online journal, 21 January 2010.
8 Winickoff, D.E., 'Genome and Nation: Iceland's Health Sector Database and its Legacy', *Innovations: Technology, Governance, Globalisation* 1, 80–105, 2006.

CHAPTER 6: THE GLOBAL COMMODIFICATION OF BIOINFORMATION

1 Fears, R. and Poste, G., 'Building Population Genetics Resources Using the UK NHS', *Science* 284, 267–8, 1999.
2 Tasmuth, T., 'The Estonian Gene Bank Project: An Overt Business Plan', *Open Democracy*, 28 May 2003.
3 Rose, H.A., 'The Rise and Fall of Umangenomics: The Model Biotech Company?' *Nature* 425, 123–4, 2003.
4 Nilson, A. and Rose, J., 'Sweden Takes Steps to Protect Tissue Banks', *Science* 286, 894, 1999.
5 Laage-Hellman, J., 'Clinical Genomics Companies and Biobanks: The Use of Biosamples in Commercial Research on the Genetics of Common Diseases', in Hansson, M.G. and Levin, M. (eds.), *Biobanks as Resources for Health*, Uppsala University, 2003, p. 71.
6 Hansson, M.G., *The Human Use of Biobanks*, Uppsala University, 2001.
7 Paulsson, U. and Friske, R., 'Agreeements Concerning Biological Material', in Hansson and Levin (eds.), *Biobanks as Resources for Health*, pp. 257–67.
8 Bibeau, G., *Le Québec transgénique*, Editions du Boréal, 2004.
9 Rustin, M., 'The New Labour Project', *Soundings* 8, 7–14, 1998.
10 Wallace, H., *Bioscience for Life? The History of UK Biobank, Electronic Medical Records in the NHS, and the Proposal for Data-sharing Without Consent*, GeneWatch, 2009.
11 Irwin, A. and Wynne, B. (eds.), *Misunderstanding Science? The Public Reconstruction of Science and Technology*, Cambridge University Press, 1996.
12 Clayton, D. and McKeigue, P.M., 'Epidemiological Methods for Studying Genes and Environmental Factors in Complex Disease', *The Lancet* 368, 1356–60, 2001.
13 Radda, G., Dexter, T.M., and Meade, T., 'The Need for Independent Scientific Peer Review of Biobank UK', *The Lancet* 359, 2282, 2002.
14 Jha, A., *Guardian*, 18 April 2006.
15 Stratton, A., 'Cameron Tries to Trigger a Big Bang', *Guardian*, 8 December 2011.
16 Bustamante, C.D., Burchard, E.G., and de la Vega, F.M., 'Genomics for the World', *Nature* 475, 163–5, 2011.
17 Weiss, L.A. et al., 'A Genome-wide Linkage and Association Scan Reveals Novel Loci for Autism', *Nature* 461, 801–8, 2009.
18 Teslovich, T.M. et al., 'Biological, Clinical and Population Relevance of 95 Loci for Blood Lipids', *Nature* 466, 709–13, 2010.
19 Couzin-Frankel, J., 'Major Heart Disease Genes Prove Elusive,' *Science* 238, 1220–1, 2010.
20 Greenwood, T.A. et al., 'Analysis of 94 Candidate Genes and 12 Endophenotypes

for Schizophrenia from the Consortium on the Genetics of Schizophrenia', *American Journal of Psychiatry* 168, 930–46, 2011.

21 Van Aken, J., Schmedders, M., Feuerstein, G., and Kollek, R., 'Prospects and Limits of Pharmacogenetics, *Amer. J. Pharmacogenomics* 3, 149–55, 2003.

22 Goldstein, D.B., 'Common Genetic Variation and Human Traits', *New England Journal of Medicine* 360, 1696–8, 2009.

23 Manolio, T.A. et al., 'Finding the Missing Heritability of Complex Diseases', *Nature* 461, 747–53, 2009.

24 Zuk, O., Hechter, E., Sunyaev, S.R., and Lander, E.S., 'The Mystery of Missing Heritability: Genetic Interactions Create Phantom Heritability', *Proceedings of the National Academy of Sciences* (US) 109, 1193–8, 2012.

25 Evans, J.P., Meslin, E.M., Marteau, T.M., and Caulfield, T., 'Deflating the Genomic Bubble.' *Science* 331, 861–2, 2011.

CHAPTER 7: THE GROWING PAINS OF REGENERATIVE MEDICINE

1 Wadman, M., 'Stem Cells Ready for Prime Time,' *Nature* 457, 516, 2009.

2 Editorial, *Guardian*, 2 February 2009.

3 Kite, M., 'Disabled People Plead for Stem Cell Research', *The Times*, 16 December 2000, quoted in Scolding, N., 'Stem-cell Therapy: Hope and Hype', *The Lancet on line*, 20 May 2005.

4 Smith, A. et al., '"No" to Ban on Stem Cell Patents', *Nature* 472, 418, 2011.

5 Habermas, J., *The Future of Human Nature*, Polity, 2003.

6 Prainsack, B., Geesink, I., and Franklin, S., 'Stem Cell Technologies 1998–2008: Controversies and Silences', *Science as Culture* 17, 351–62, 2008.

7 International Stem Cell Forum Ethics Working Party, *Science* 312, 366–7, 2006.

8 Needham, J., *A History of Embryology*, Cambridge University Press, 1934.

9 Wolpert, L., *The Triumph of the Embryo*, Oxford University Press, 1991.

10 Takashi, K., Tanabe, K., Ohnuki, M., Narita, M., Ichisaka, T., Tomoda, K., and Yamanaka, S., 'Induction of Pluripotent Stem Cells from Adult Human Fibroblasts by Defined Factors', *Cell* 131, 1–12, 2007.

11 Gage, F.H., Dunnett, S.B., Stenevi, U., and Bjorklund, A., 'Aged Rats: Recovery of Motor Impairments by Intrastriatal Nigral Grafts,' *Science* 221, 966–9, 1983.

12 Stein, D.G. and Glasier, M.M., 'Fetal Brain Tissue Grafts as Therapy for Brain Dysfunctions: Some Practical and Theoretical Issues', *Behavioral and Brain Sciences* 18, 36–45, 1995.

13 Kolata, G., 'Brain Grafting Work Shows Promise,' *Science* 221, 1277, 1983.

14 Madrazo, I., Drucker-Colin, V., Diaz, V., Martinez-Mata, J., Torres, C., and Becerril, J.J., 'Open Microsurgical Autograft of Adrenal Medulla to the Right Caudate Nucleus in Two Patients With Intractable Parkinson's Disease', *New England Journal of Medicine* 316, 831–4, 1987.

15 Quoted in Lewin, R., 'Brain Grafts Benefit Parkinson's Patients,' *Science* 236, 149, 1987.

16 Sladek, J. and Shoulson, I., 'Neural Transplantation: A Call for Patience Rather Than Patients, *Science* 240, 1386–8, 1988.

17 Thomson, J., Itskovitz-Eldor, J., Shapiro, S.S., Waknitz, M.A., Swiergiel, J.J., Marshall, V.S., and Jones, J.M., 'Embryonic Stem Cell Lines Derived From Human Blastocysts', *Science* 282, 1145–7, 1998.

18 Gearhart, J., 'New Potential for Human Embryonic Stem Cells, *Science* 282, 1061–2, 1998.

19 Hwang, W.S. et al., 'Evidence of a Pluripotent Human Embryonic Stem Cell Line Derived From a Cloned Blastocyst', *Science* 303, 1669–74, 2004.

20 Vogel, G., 'Scientists Take Step Towards Therapeutic Cloning', *Science* 303, 937, 2004.

21 Kennedy, D., 'Stem Cells Redux', *Science*, 303, 1581, 2004.

22 Hwang, W.S., et al., 'Patient-specific Embryonic Stem Cells Derived From Human SCNT Blastocysts', *Science* 308, 1777–83, 2004.

23 *HFEA Welcomes Korean Scientists' Stem Cell Breakthrough*, HFEA press release, 12 February 2004.

24 Sciencemediacentre.org/press_releases/04-02-12_stemcells.htm consulted 09/05/08.

25 Vogel, 'Scientists Take Step Towards Therapeutic Cloning'.

26 Cyranoski, D., 'Crunch Time for Korea's Cloning', *Nature* 429, 12–14, 2004.

27 Cyranoski, D., 'Korean Stem-cell Crisis Deepens', *Nature* 438, 405, 2005.

28 Human Embryology and Fertilisation Authority, *Hybrids and Chimeras*, 2007.

29 Academy of Medical Sciences, *Animals Containing Human Material*, 2011.

30 Dolgin, E., 'Flaw in Induced Stem Cell Model', *Nature* 470, 13, 2011; Apostolou, E. and Hochedlinger, K., 'iPS Cells Under Attack', *Nature* 474, 165–6, 2011.

31 Kiatpongsan, S. and Sipp, D., 'Monitoring and Regulating Offshore Stem Cell Clinics', *Science* 323, 1564–5, 2009.

32 Murdoch, C.E. and Scott, C.T., 'Stem Cell Tourism and the Power of Hope', *American Journal of Bioethics* 10, 16–23, 2010.

CHAPTER 8: THE IRRESISTIBLE RISE OF THE NEUROTECHNOSCIENCES

1 Lynch, Z., Videoconference, March 2012.

2 Crick, F., *The Astonishing Hypothesis*, Simon and Schuster, 1994.

3 Kandel, E.R., *In Search of Memory: The Emergence of a New Science of Mind*, Norton, 2007.

4 Churchland, P., *Neurophilosophy: Towards a Unified Science of the Mind-Brain*, Bradford Books, 1986.

5 Solms, M. and Turnbull, O., *The Brain and the Inner World: An Introduction to the Neuroscience of Subjective Experience*, Other Press, 2002.

6 Edelman, G.M., *Wider Than the Sky: The Phenomenal Gift of Consciousness*, Allen Lane, 2004.

7 Libet, B., *Mind-Time: The Temporal Factor in Consciousness*, Harvard University Press, 2004.

8 Wittchen, H.U. et al., 'The Size and Burden of Mental Disorders and Other Disorders of the Brain in Europe 2010', *European Journal of Neuropsychopharmacology* 21, 665–79, 2011.

9 Rose, S., *The Making of Memory*, Vintage, 2003.

10 Vul, E., Harris, C., Winkielman, P., and Pashler, H., 'Puzzlingly High Correlations in fMRI Studies of Emotion, Personality and Social Cognition', *Perspectives on Psychological Science* 4, 274–90, 2009.
11 Bennett, C.M., Baird, A.A., Miller, M.B., and Wolford, G.L., 'Neural Correlates of Interspecies Perspective Taking in the Post-mortem Atlantic Salmon: An Argument for Multiple Comparisons Correction, *Society for Neuroscience Abstracts*, 2009.
12 Hsu, M., Anen, C., and Quartz, S.R., 'The Right and the Good: Distributive Justice and the Neural Encoding of Equity and Efficiency', *Science* 320, 1092–5, 2008.
13 Outside the US, mental and nervous system diseases are included in WHO's International Classification of Diseases, the ICD.
14 Kramer, P.D., *Listening to Prozac*, Penguin, 1993.
15 Whittaker, R., *Anatomy of an Epidemic: Could Psychiatric Drugs be Fueling a Mental Illness Epidemic?*, Random House, 2010.
16 Healy, D., 'Psychopharmacology at the Interface Between the Market and the New Biology', in Rees, D. and Rose, S. (eds.), *The New Brain Sciences: Perils and Prospects*, Cambridge University Press, 2004, pp. 232–48.
17 Smith, R., Talk at EMBO Conference on Understanding Mental Disorder, November 2011.
18 Healy, D., *Let Them Eat Prozac*, Lorimer, 2003.
19 Cornwell, J., 'The Prozac Story', in Rees and Rose (eds.), *The New Brain Sciences*, pp. 223–31.
20 Editorial, 'Depressing Research', *The Lancet* 363, 1335, 2004.
21 Maher, B., 'The Case of the Missing Heritability', *Nature* 456, 21, 2008.
22 National Research Council of the National Academies, *Emerging Cognitive Neuroscience and Related Technologies*, 2010.
23 Glueck, S. and Glueck, E., *Unraveling Juvenile Delinquency*, The Commonwealth Fund, 1950.
24 Brennan, P.A., Mednick, S.A., and Jacobsen, B., 'Assessing the Role of Genetics in Crime Using Adoption Cohorts', in Bock, G.R. and Goode, J.A. (eds.), *Genetics of Criminal and Anti-social Behaviour*, Ciba Foundation Symposium 194, Wiley, 1996, pp. 115–22.
25 Brunner, H.G., Nelen, M., Breakfield, X.O., Ropers, H.H., and van Oost, B.A., 'Abnormal Behavior Associated with a Point Mutation in the Structural Gene for Monoamine Oxidase A', *Science* 262, 578–80, 1993.
26 Kelso, J.R., 'Behavioural Neuroscience: A Gene for Impulsivity', *Nature* 468, 1049–50, 2010.
27 Caspi, A. et al., 'Role of Genotype in the Cycle of Violence in Maltreated Children', *Science* 297, 851–4, 2002.
28 Caspi, A. et al., 'Influence of Life Stress on Depression: Moderation by a Polymorphism in the 5-HTT Gene', *Science* 301, 386–9, 2003.
29 Cooper, P., 'Education in the Age of Ritalin', in Rees and Rose (eds.), *The New Brain Sciences*, pp. 249–62.
30 Better perhaps than being arrested by an armed police officer patrolling the classrooms, as is apparently becoming routine in some US states (*Guardian*, 10 January 2012).
31 Motzkin, J.C., Newman, J.P., Kiehl, K.A., and Koenigs, M., 'Reduced Prefrontal Connectivity in Psychopathy, *Journal of Neuroscience* 31, 17348–57, 2011.

32 University Of Southern California, 'USC Study Finds Faulty Wiring in Psychopaths', *ScienceDaily*, 11 March 2004.
33 Editorial, 'Deceiving the Law', *Nature Neuroscience*, 11, 1231, 2008.
34 National Research Council of the National Academies, *Emerging Cognitive Neuroscience and Related Technologies*, 2010.
35 Sedley, S., 'Responsibility and the Law', in Rees and Rose (eds.), *The New Brain Sciences*, pp. 123–30.
36 Royal Society, *Neuroscience and the Law, Brain Waves Module 4*, 2012.
37 Wilson, E.O., *On Human Nature*, Harvard University Press, 2004.
38 Jencks, C., 'EP Phone Home', in Rose, H. and Rose, S. (eds.), *Alas Poor Darwin: Arguments Against Evolutionary Psychology*, Cape, pp. 33–54, 2000.

CHAPTER 9: PROMETHEAN PROMISES: WHO BENEFITS?

1 Sinden, J., welcoming Hwang's claims at the Science Media Centre in 2004.
2 Braude, P., quoted in *Mail Online*, 7 December 2011.
3 Furcht, L. and Hoffman, L., *The Stem Cell Dilemma: The Scientific Breakthroughs, Ethical Concerns, Political Tensions, and Hope Surrounding Stem Cell Research*, Arcade, 2011.
4 Ascoli, G., quoted in 'Searching for the "Holy Grail" of Neuroscience: Mason Scientist Aims to Construct a Virtual Brain', *The Mason Gazette*, 5 February 2007.
5 Chain, D., 'Intellect Neurosciences Chases Holy Grail of Alzheimer's: A Preventive Vaccine, *Fierce Vaccines*, 10 April 2012.
6 Liberles, S., quoted on bostonblog, 'Harvard Discovers the Holy Grail of Neuroscience', 27 June 2011, blogs.nature.com/boston.
7 Quoted on BookRags.com, Summary of the Human Genome Project.
8 Marks, J., *What it Means to be 98% Chimpanzee*, University of Chicago Press.
9 Evans J.P., Meslin, E.M., Marteau, T.M., and Caulfield, T., 'Deflating the Genomic Bubble', *Science* 331, 861–2, 2011.
10 Collins, F.S., *The Language of Life: DNA and the Revolution in Personalised Medicine*, Harper, 2010.
11 Novas, C. and Rose, N., 'Genetic Risk and the Birth of the Somatic Individual', *Economy and Society* 29:4, 485–513, 2000.
12 Cholesterol Treatment Trialists' (CTT) Collaborators. 'The Effects of Lowering LDL Cholesterol With Statin Therapy in People at Low Risk of Vascular Disease: Meta-analysis of Individual Data From 27 Randomised Trials', *The Lancet*, Early Online Publication, 17 May 2012; DOI:10.1016/S0140- 6736(12)60367-5.
13 Healy, D., 'Psychopharmacology at the Interface Between the Market and the New Biology', in Rees, D. and Rose, S. (eds.), *The New Brain Sciences: Prospects and Perils*, Cambridge University Press, 2004, pp. 234–49.
14 Kahn, J., 'Race in a Bottle', *Scientific American* 297:2, 40–5, 2007.
15 Breggin, P. and Cohen, D., *Your Drug May Be Your Problem: How and Why to Stop Taking Psychiatric Medications*, Perseus, 2007.
16 *Guardian*, 27 April 2012.
17 Monbiot, G., 'Britain's Shadow Government: Unelected, Unbalanced and Unaccountable', *Guardian*, 12 March 2012.

18 Abate, T., *San Francisco Chronicle*, 13 August 2001.

19 Vines, G., *Raging Hormones: Do They Rule Our Lives?* Virago, 1993.

20 Oudshoorn, N. *Beyond the Natural Body. An Archeology of Sex Hormones*, Routledge, 1994.

21 Baron-Cohen, S., *The Essential Difference: The Truth About the Male and Female Brain*, Basic Books, 2003.

22 Jordan-Young, R., *Brainstorm: The Flaws in the Science of Sex Differences*, Harvard University Press, 2010.

23 Fine, C., *Delusions of Gender: The Real Science Behind Sex Differences*, Icon, 2010.

24 Caplan, A., 'If It's Broken, Shouldn't It Be Fixed? Informed Consent and Initial Clinical Trials of Gene Therapy', *Human Gene Therapy* 19, 5–6, 2008.

25 Conrad, E. and de Vries, R., 'Field of Dreams: A Social History of Neuroethics', *Sociological Reflections on the Neurosciences*, Advances in Medical Sociology, Emerald Publishing, 13, 299–324, 2011.

26 Pilcher, H., 'Dial E for Ethics', *Nature* 440, 1104–5, 2006.

27 Bauman, Z., *Post-Modern Ethics*, Blackwell, 1993.

28 Rapp, R., *Testing Women, Testing the Fetus*, Routledge, 2000.

29 Latour, B. and Woolgar, S., *Laboratory Life: The Social Construction of Scientific Facts*, Sage, 1979.

30 Gross, P. and Levitt, N., *The Higher Superstition: The Academic Left and its Quarrels With Science*, Johns Hopkins University Press, 1994.

31 Palsson, G., 'The Rise and Fall of a Biobank: The Case of Iceland', in Gottweis, H. and Petersen, A. (eds.), *Biobanks: Governance in Comparative Perspective*, Routledge, 2008, pp. 41–55.

32 Titmuss, R., *The Gift Relationship: From Human Blood to Social Policy*, 1970, reprinted by The New Press, 1997.

33 Leach, E., Review of the Gift Relationship, *New Society*, 12 January 1971.

34 Nuffield Council on Bioethics, *Human Bodies: Donation for Medicine and Research*, 2011.

35 Murdoch, A., quoted in Wellcome Trust press release, *Techniques to prevent transmission of mitochondrial diseases to be assessed in new £5.8 million Wellcome Trust centre,* Wellcome Trust, 19 January 2012

36 Nuffield Council on Bioethics, *Solidarity: Reflections on an Emerging Concept in Bioethics*, 2012.

Index